JAMES MAY'S 20TH CENTURY

BBC

JAMES MAY AND PHIL DOLLING

HODDER & STOUGHTON

First published in Great Britain in 2007 by Hodder & Stoughton

An Hachette Livre UK company

2

Copyright © James May and Phil Dolling 2007

The right of James May and Phil Dolling to be identified as the
Authors of the Work has been asserted by them in accordance
with the Copyright, Designs and Patents Act 1988.

By arrangement with the BBC
The BBC logo is a registered trademark of the British Broadcasting
Corporation and is used under licence
BBC logo © BBC 1996

A CIP catalogue record for this title is available from
the British Library

Hardback ISBN 978 0 340 95090 6
Trade Paperback ISBN 978 0 340 95091 3

Design by Rose
Typeset in Futura

Printed and bound in Great Britain by Butler and Tanner

Hodder & Stoughton policy is to use papers that are natural,
renewable and recyclable products and made from wood grown
in sustainable forests. The logging and manufacturing processes
are expected to conform to the environmental regulations of the
country of origin.

Hodder & Stoughton Ltd
338 Euston Road
London NW1 3BH

www.hodder.co.uk

TO MEN IN SHEDS, EVERYWHERE.

CONTENTS

INTRODUCTION

Around 1.6 million years ago our distant ancestors, Homo ergaster, created a cutting tool, a simple hand axe in the shape of a tear drop. It worked very well; so well, in fact, that they and their descendants barely altered it for the next million years.

A million years without progress is incomprehensible. If all innovation happened at such a glacial pace, inhabitants of the Earth in 1,002,007 would still be driving around in the new Fiat Panda and still listening to James Blunt.

Fortunately, something happened in the course of human evolution and encouraged ingenuity. Some suggest it was the mastery of fire that ultimately led to the white heat of technology. Others believe it was something more abstract, the development of consciousness. I can't help wondering, though, if it was the invention of the shed.

Eventually, successful technology leads to huge corporations, research laboratories and great sprawling factories employing thousands of people between them. But the moment of inspiration is invariably a lonely experience in a shed of some sort. The Wright Brothers built a shed at Kitty Hawk and assembled the Flyer there. It was in a shed that Karl and Bertha Benz stood mesmerised by the faltering thud of their internal combustion engine. John Logie Baird developed his television system in an attic, which is nothing more than a shed structure on top of another building.

It may be simply that inventors are usually poor, notoriously bereft of business acumen and generally misunderstood, and so are forced by circumstances into sheds. But it could be that the shed environment, and some unwritten convention that says it is a place of deep thought that is not subject to the restrictive tenets of normal life, provides a unique form of inspiration. Sometimes, the shed simply grows with the development of the new science: Britain's airship industry existed entirely in giant sheds, and the rocket that took men to the Moon, the Saturn V, was built in NASA's VAB, the Vehicle Assembly Building in Florida and the biggest and greatest shed of all time.

Whatever it is that promotes innovation, the 20th century enjoyed it in abundance. In the space of 100 years the sky filled with aeroplanes, humans set foot on another planet, the globe was shrunk by radio and television, the common man was liberated by the car and the motorcycle, we swam with exotic undersea creatures purely for recreation and we listened out for signs of life from the other side of the universe. And as we expanded our horizons outwards, we also turned our curiosity inwards to investigate the nature of ourselves, learning in the process that there is as much to discover in our own bodies as there is in the vastness of space.

But this leads to a problem; what to include in a book like this and the television series that goes with it. Not everything, certainly, because there is so much. So this is, unashamedly, a personal selection, the inventions and discoveries that best capture the mood of the most exciting century in history. Of course not all innovation is for the better and some, such as the mobile phone and the atomic bomb, threaten the very fabric of society. But the net result of progress is magnificent and should – must – be celebrated.

The alternative is stark: the only tool would be a stone hand axe just like the one that belonged to your father x 40,000.

SHRINKING THE WORLD
AVIATION
CARS
TELEVISION
COMPUTERS

1

THE GREAT DEPARTURE

Just one of a number of Concordes, although no-one ever saw it that way.

Aviation enthusiasts have always bought aeroplane picture books. At any air show they are racked up for the delectation of those people with unfeasibly long camera lenses, and sporting titles such as *The Boeing 757 colour portfolio.*

But *Concorde*, by the photographer Wolfgang Tillmans, is different. It is still merely a collection of pictures – snapshots, really – taken around London and showing the aeroplane soon after take-off and on its final approach. The difference is that Tillmans's book found its way on to the coffee tables of people who have never been to Farnborough and who would barely look skywards if a jumbo flew over.

It helps that Tillmans is really an artist, and that *Concorde* is considered an art book. It helps also that its subject-matter is as unutterably beautiful as a Raphael virgin. Even so, his book would not enjoy the unique place it occupies amongst aviation literature if Concorde did not exert such a hold over the consciousness of the people.

Consider this: Concorde is an aeroplane type, of which twenty were built, yet no one ever said, 'There goes a Concorde.' It was always simply 'Concorde', an entity rather than a mere industrial artefact. No other manufactured product enjoys this distinction of existing in number yet being perceived as unique.

Concorde was totemic, iconic, and everything else that has been said about it a million times; an airborne monument to the white heat of technology and a testimony to the pace of progress in the century that produced it. The same century gave us the very first powered aeroplane, in 1903, but less than a lifetime later, in 1969, gave us one that could fly at twice the speed of sound, so high that the sky turned black, and while its hundred passengers dined in pressurised and air-conditioned comfort.

Today, Concorde G-BBDG – claimed to be the fastest of the fleet – resides at the Brooklands Museum as a tourist attraction. Visitors can climb the steps and settle into the snug seats of its surprisingly narrow fuselage,

'ONLY AN AIR SPEED INDICATOR READING TO MACH 2.2 DISTINGUISHES THE COCKPIT OF CONCORDE FROM THAT OF A COLD WAR-ERA V-BOMBER'

as they would have done if they were ever rich or important enough to fly on it for real. The bulkhead-mounted speed readout will be dutifully made to flicker into life and display the magical Mach 2 figure that the original occupants paid so much to experience. Even on the ground, the Concorde experience is quite magical.

But slip past the 'no entry' sign and into the cockpit and the mystique of Concorde begins to show a few cracks. This is a sixties aircraft, conceived before the digital age and, while the seat fabric, lavatories and galley service were constantly improved, the costs and complications of certification meant that the flight deck went largely unchanged. So there is no fly-by-wire joystick control, no 'glass cockpit' computer screen and, frankly, no space to move. Instead there is a huge and weighty control yoke, serried ranks of clunky switches, massive levers for engine controls and an incomprehensible array of analogue instruments set into the chipped and faded grey expanse of its panel. Only an airspeed indicator reading to Mach 2.2 distinguishes the cockpit of Concorde from that of a Cold War-era V-bomber.

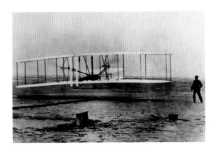

Orville Wright takes to the air, 17 December 1903. Older brother Wilbur runs alongside.

Left:
The hideously cramped Concorde flight-deck, as seen from the flight engineer's seat.

The pilot's seat of Delta Golf is a good place for an aviation reality check. The debate about the value of Concorde still rages. It was a great morale booster, a symbol of national pride for the countries that built it, an inspiration to all industry. It is said that more aerodynamic theory was thrashed out in the design of Concorde's delta wing than in all the preceding years of heavier-than-air flight. But it was hideously expensive, it was considered anti-social by many, it never entered widespread service with the world's airlines, as its creators hoped, and in the era of modern telecommunications it began to look painfully slow as a means of linking London and New York. It remains the only supersonic airliner to have entered regular service, but belongs to an age that has passed.

Seated in the now silent shrine to Concorde's captain and first officer, staring through the clutter of struts, seals and perspex that form Delta Golf's raised 'droop nose', it becomes easier to accept that the most dazzling commercial aircraft in history might just be its most irrelevant.

Manned flight is a technology that must have been envisaged long before it was achieved. There is, after all, a precedent in nature. As soon as humans walked upright and turned their eyes upwards, they saw birds doing it and must have sensed it was possible.

But that's not to say it was easy. Attempts at flight are as old as civilisation itself, but it was not until the first few years of the twentieth century that the aeroplane as we might recognise it took to the air, courtesy of a couple of American bicycle-makers called the Wright brothers (only five Wright-built bicycles are known to have survived).

On the morning of 17 December 1903, at a remote and desolate beach near Kitty Hawk in North Carolina, Orville Wright kept their confidently named 'Flyer' in the air for twelve seconds. Owing to a hefty headwind of almost 25 knots, he covered a distance of just 120 feet, so the world's first aeroplane trip could be replicated in its entirety within the shadow of a Boeing 737.

Orville, in turn with his elder brother Wilbur, made several more flights that day, the longest covering 852 feet and lasting almost a minute. These were not the fortuitous 'hops' that had been achieved in some other early experiments. This was flight, and the historic day was brought to an end only when the machine was damaged in a minor crash.

Today, the original Flyer is in the Smithsonian Institution in Washington DC and is one of its most prized exhibits. But it wasn't placed there immediately. For years a debate raged over who was first – the Wright brothers, or their rival, Samuel P. Langley (secretary of the Smithsonian, no less), whose 'Aerodrome' crashed into the Potomac river earlier in 1903 but was exhibited in the institution as the first machine *capable* of sustained flight. The argument continued until 1948 when, following the intervention of Charles Lindburgh, the Flyer was moved from London's Science Museum and given its rightful place.

There have been other contenders, too, such as the French pioneer Felix du Temple de la Croix, who claimed to have flown his steam-powered

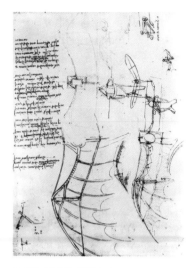

Top:
Plenty of people sketched birds in the 16th century, but Leonardo da Vinci worked out how they flew.

Bottom:
Otto Lillenthal airborne in one of his gliders in 1896. He died later that year in an early aviation incident.

aeroplane as early as 1874. Another Frenchman, Clement Adler, is purported to have flown his steam monoplane a distance of 165 feet in 1890. The German Gustav Weisskopf is said to have flown half a mile in 1901 and, astonishingly, seven miles in 1902. Cynics have suggested that the principal achievement of the Wright brothers was in remembering, unlike any of their rivals, to pack a camera.

But the Wright brothers, instinctive practical engineers who built bicycles for a living but also dabbled in engines and printing presses, deserve the honour. What they actually achieved, in the strict language of the historian, was the world's first sustained, powered and controlled flight of a heavier-than-air machine. And control was the big issue.

Looking at the Wrights' Flyer today, it seems that they got the basic layout of the aeroplane slightly wrong. What we would now think of as the tailplane is in front of the main wings, with only the rudder behind. The pilot was also required to adopt a prone posture. But the system of control they devised is essentially the one that survives to this day (curiously, their *means* of control was a truly counter-intuitive series of levers, the arrangement of which was rendered almost immediately obsolete). One detail difference is in the control of roll, which the Wrights achieved by warping the whole wing of the machine, an idea inspired by observations of bird flight (as, indeed, it was in the sketches of da Vinci). Today, that job is done by ailerons – smaller hinged surfaces set into the wing – but the principle is broadly the same. Prior to the Wrights, it was generally believed that an aeroplane would be steered by shifting the pilot's body weight, as a hang glider is. (A legal battle with rival aviation pioneer Glenn Curtis over the patent rights to the aileron principle was just one of several in which the Wrights became embroiled.)

It would be wrong, though, to imagine that aviation pioneers spent centuries flapping around in Icarus-like feathered bird suits only for the Wrights to step in and solve the whole conundrum at a stroke. Much valuable work had already been done prior to their historic first flight, including, in fact, quite a bit of flying.

In England, the Yorkshire baronet Sir George Cayley (1773–1857) worked tirelessly on theories of flight, and a sketch of his dated as early as 1799 shows the basis of a fixed-wing aeroplane and some of the aerodynamic forces acting upon it. In Germany, the mechanical engineer Otto Lillenthal and his English pupil Percy Pilcher had made many successful manned glider flights before their deaths, in flying accidents, in 1896 and 1899 respectively. Models had been flown – noticeably by the Wrights' eventual detractor, Samuel Langley – and the American Octave Chanute, a tireless champion of the quest for flight, produced a learned work entitled *Progress in Flying Machines*. The Wrights knew of all this, and acknowledged its importance in their own work. They, too, began their experiments with gliders and, using their own miniature wind-tunnel, sorted the issue of control before they took the bold step of adding power from a 12 hp petrol engine (which, inevitably, they designed and built themselves).

The Wrights were modest and secretive about their work. Although scores of people saw their miraculous flying machine in the air as they conducted further trials near their home in Dayton, Ohio, it was not until 1907 that a flight of more than a minute's duration was achieved in Europe, by Henri Farman in a Voisin biplane. But once fired, enthusiasm for aviation in Europe was such that the Wrights' aeroplanes were soon outmoded, and it was the French who took the lead. The year 1909 saw the first of what we would now call an air display, the Grande Semaine d'Aviation de la Champagne at Rheims, and in the same year it was another Frenchman, Louis Blériot, who conquered the second great aviation milestone – a crossing of the English Channel by air.

On 25 July, starting from the coast at Calais, he flew the 23.5 miles to the Dover clifftops in a monoplane of his own design, crash-landing near the castle but emerging unscathed to enjoy a hero's welcome by the British. It is thought that it was only the cooling effect of light drizzle on Blériot's struggling Anzani engine that prevented him from losing power and ending his bid in the sea, but no matter. The aeroplane could be seen to have arrived and, portentously, the far-sighted recognised that maritime

Desperate to maintain height, the French airman discarded his roses.

Overleaf:
The Wright Brothers founded a flying school in Pau in 1909, but within a few years the French had left them behind.

Louis Blériot rejoices deliriously following the success of his cross-channel flight.

Opposite page
Top:
Alcock and Brown's Vickers Vimy leaves Newfoundland on the world's first long-haul flight, 14 June 1919.

Bottom:
One of Brown's hand-written notes to his pilot. Conversation was impossible over the racket of the engines.

isolation could no longer be regarded as a guarantee of safety from aggressors. For his feat, Blériot received the £1,000 prize offered by the *Daily Mail*; the cash and his new-found reputation helped him establish a highly successful aeroplane business.

By 1913, the concept of a passenger-carrying machine had already been demonstrated by the genius Igor Sikorsky with his massive ninety-two-foot-long Bolshoi airliner. It was the first four-engined aeroplane to fly, the largest flying machine of its time, and the first to feature an enclosed passenger cabin. On 2 August of that year it made a flight of 1 hour 54 minutes with eight people on board. The world was about to witness a great dawn when it was plunged into a war, among the first casualties of which was the promise of commercial air travel.

But in a roundabout way, the war that threatened to destroy the world ultimately played a vital role in shrinking it. From 1914 to 1918 aviation technology progressed by the leaps and bounds it always makes during wartime. Most notably, aero engines were developed to vastly superior levels of power and reliability, leading to parallel improvements in the performance, range and dependability of the aeroplanes to which they were fitted. Aeroplanes at the beginning of the war were fitted with the sort of engine that had taken Blériot over the Channel. By 1918 Rolls-Royce was building the Eagle VIII, a magnificent V12 monster producing 360 hp.

Two Eagle VIII engines powered the Vickers Vimy, a biplane of sixty-eight-foot wingspan designed to be Britain's heavy bomber, but arriving too late to play an effective role in the war. It was with a Vimy – in essence an item of war surplus – that two Englishmen, John Alcock and Arthur Whitten Brown, made their attempt on the next great milestone in aviation, a non-stop crossing of the Atlantic.

Both men were veterans of the first air war, and both had been shot down and taken prisoner. At the suggestion of Vickers, Alcock, as pilot, and Brown, as navigator, teamed up for the treacherous transatlantic bid and a chance to win the £10,000 prize on offer for the first successful crossing, again put up by the *Daily Mail*.

To take advantage of the tailwind that can still knock a couple of hours off the return flight time from New York to London, the Vimy was dismantled and shipped to Newfoundland. It was reassembled in a suitable field near St John's and loaded with 870 gallons of fuel, forty gallons of oil, a few meagre provisions and two stuffed toy cats as lucky mascots. On 14 June 1919, Alcock and Brown lumbered into the air, missing surrounding trees by inches, and set course for the British Isles. As they thundered over St John's Harbour at 1,000 feet, vessels sounded their sirens in final farewell.

It seems likely that some of the in-flight escapades of the two pioneers were somewhat exaggerated by press reports of the time. But even without the journalistic licence, the journey was harrowing enough. The Vimy could cruise at an airspeed of around 90 mph – a speed comfortably surpassed today by a Fiat Panda. Their radio packed up after three hours and they flew into desperate weather. Most of the journey was made in cloud or a fog so thick that even the engines, a few feet away on the wings, were obscured. Before nightfall one brief break in the weather allowed a sun-sighting, confirming that the Vimy was only a few miles off course and that Brown was indeed a superlative navigator.

The racket from the engines was too great for conversation, so the pair communicated with written notes. The battery-operated system for their heated suits failed, as did the altimeter. At one point, the Vimy fell into a disorientating spiral dive in cloud, Alcock recovering control only twenty feet above the sea. Both men reported tasting salt spray on their lips.

Miraculously, after fifteen hours in the air, Alcock and Brown spotted the Irish coast in the distance. They flew low on arrival, looking for a suitable landing ground. A green meadow looked ideal, not least because the locals appeared to be waving them in. Unfortunately, the gestures were a warning that the 'meadow' was in fact a bog, Derrygimla Moor. After 1,890 miles and 16 hours of knife-edge airmanship, Alcock and Brown's Vimy came to grief, nosed over in the mud with its wings and fuselage badly damaged but with its crew unhurt. When Alcock announced that they had just flown in from America, the assembled crowd laughed at the absurdity of the suggestion.

Fame, money and knighthoods followed, and seem more than deserved when Alcock and Brown's original Vimy is viewed at close quarters in London's Science Museum. It seems a fabulously primitive machine, apparently the combined work of a shed-builder and a sail-maker. Popular history often recognises Charles Lindbergh's flight from New York to Paris in 1927 as the first transatlantic crossing; it was merely the first *solo* crossing.[1] Lindbergh's achievement is not to be diminished, but Alcock and Brown's was more significant and more momentous. It remains the most epic flight in aviation history and the equal, for daring and audacity, of the Apollo 11 moonshot of fifty years later.

It certainly stoked the enthusiasm for commercial air travel that had been kindled with Sikorsky's Bolshoi. In the same year as the Atlantic

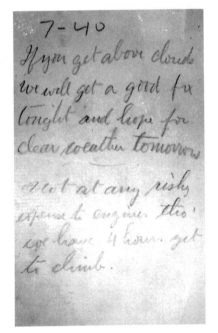

The interior of a Handley Page O/400 airliner, a converted bomber type. It was clearly designed on a colonial verandah to look like one.

Left:
Imperial Airways tempts passengers away from the ocean-going liners with its range of modern aeroplanes.

crossing, and just a month later, Britain's first airline, the short-lived Aircraft Transport and Travel, began operations with another converted First World War bomber, the single-engined De Havilland 4A. Old bombers were the stuff of pioneer airlines.[2]

The Dutch national carrier KLM was also founded in 1919, and remains the oldest airline still operating under its original name. Its first flight, the following year, took two English passengers from London to Schipol. As with many other fledgling European airlines, it was founded with a view to maintaining good communications with distant colonial possessions. The UK's long-haul flag-carrier, Imperial Airways, was set up to compete with the Europeans in 1924. It enjoys the distinction of being the first airline to show an in-flight film, *The Lost World*, between London and Paris in 1925. It was to evolve, by 1939, into the British Overseas Airways Corporation after merging with the then private British Airways Ltd.[3]

Imperial would eventually use ruggedly handsome flying boats, such as the Short Brothers' Ensign and Empire for routes to Africa, and the precarious-looking Handley Page HP42 for the overland route to India. But in its early days it was forced to rely, like everyone else, on thinly modified ex-service aircraft in which bomb racks and gun positions had given way to a few creaky conservatory chairs. This was an era in which merely being a passenger was a pioneering activity, in which the most intrepid could forgo the luxury of wickerwork and join the pilot in the open cockpit. Arrival at any specific landing place could not be guaranteed, and turbulence at the enforced low altitudes could be severe.

So it is hardly surprising that, even in the 1920s, there were plenty who doubted the aeroplane's long-term future, and especially where long-haul was concerned. There was, after all, an alternative, in the form of the airship.

By the 1920s, the airship was already an established means of travel. Deutsche Luftschiffahrts AG, or Delag, was founded in 1909, its purpose to operate passenger flights using Count Ferdinand von Zeppelin's incredible craft. By the outbreak of the First World War Delag had logged 1,500 flights covering 100,000 miles and carrying 35,000 passengers, all without mishap. This, after all, was the world's first true airline.

Despite a spot of bad PR during the war, when Zeppelins were used to bomb London, the German airship industry survived (with some help from American funding) and continued to stun the world. In 1924 the Zeppelin LZ126, ordered by America as a war reparation, was flown across the Atlantic to a rapturous reception. President Coolidge called it 'an angel of peace' before assuming possession and renaming it the USS *Los Angeles*. The next Zeppelin, the LZ127 *Graff Zeppelin*, did much to cement the popularity of the airship, circumnavigating the globe in 1929 and flying a team of researchers to the Arctic in 1931.

In Britain, the peacetime potential of the airship had not gone unrecognised and Britain would therefore build the biggest and greatest. Two huge designs were begun in the mid-twenties; one, the R100, was a private venture by Vickers, built at Howden in Yorkshire (the chief engineer was Barnes Wallis, later of bouncing-bomb fame). The other, R101, was a

'APART FROM ITS USE AS AN ADVERTISING BLIMP... THE AIRSHIP HAS BEEN DISMISSED AS A RED HERRING OF AVIATION'

government job, built at the Royal Airship Works at Cardington.

Today, all that survives of the two projects is the pair of massive assembly sheds at Cardington; one restored and in use as a film studio, the other decaying, windswept, largely untouched, and deeply haunted by the spirits of the men who had this magnificent and dignified vision of shrinking the world with lighter-than-air flight. Apart from its continued use on a smaller scale as an advertising blimp or a floating television camera platform, the airship has been dismissed as a red herring of aviation.

But at the time, the argument for the airship was highly convincing. An envisaged 'voyage' from England to India in the R101 would take five days. The same trip on Imperial's aeroplane would take eight days and would involve twenty-one stops. The quickest sea passage took four weeks.

And above all, airships were luxurious. Instead of the buffeting and claustrophobia of the aeroplane, the R101 would offer the appointments of a first-class hotel: sleeping quarters, bars, dining rooms, promenades and even a smoking room. It was in the smoking room that the nervous

In three days to South America! A ship could have taken three weeks.

James May's 20th Century: The Great Departure

The R101 Airship in its shed at Cardington. Not for long, as it would turn out.

passenger might perceive the one potential drawback of being suspended under several million cubic feet of highly volatile hydrogen gas. The room itself was lined with asbestos, and the only lighter allowed on board was chained to its table.

The vastness of the Cardington sheds is a clue to the volume of gas needed to transport a relatively small number of people; their size the only remaining clue to how awe-inspiring these lightweight leviathans of the air must have looked. R101 was almost 800 feet long. A ship indeed.[4] Given its volume – over 5.5 million cubic feet – the use of hydrogen as its means of levitation seems like pure lunacy. But there were reasons: helium, the alternative, was harder to come by and very expensive. Hydrogen could be produced on-site by a relatively simple process.

When R101 crashed on its maiden voyage to India on 5 October 1930, during stormy weather over France, the impact itself was hardly dramatic. It is reckoned to have been travelling at just 13 mph, the ideal speed for a perfect landing, as a survivor observed. It settled with a light bump. Sadly, the fire that seems to have been caused by a hot engine coming into contact with the ruptured gasbags killed all but six of the fifty-four on board. Among the dead were seventeen high-ranking VIPs. It was a national disaster, a *Titanic* of the air. R100 was scrapped and the British airship industry faded away.

In a perverse way, the crash of the R101 almost prevented the most famous airship disaster of all time, that of the *Hindenburg*. The Zeppelin company's boss, the brilliant and pacifistic Dr Hugo Eckener, had been alarmed by the speed at which fire had engulfed the British airship and decided that his next Zeppelin should be filled with inert helium.

Unfortunately, by the time it was launched in 1936 the Nazis were in power. Although they saw no military use for airships, they were keen to exploit the propaganda benefits of the much-admired Zeppelins – hence the disturbing swastikas on the *Hindenburg*'s fin. Eckener himself had named the ship after the former German president Paul von Hindenburg to pre-empt any attempt by the party to name it after Hitler.

But the Nazis' obvious warmongering meant that there was an embargo on helium, which only the US could supply in sufficient quantities. So hydrogen it was, although as a precaution the gas was tainted with crushed garlic so that leaks would be easy to detect.

The *Hindenburg* disaster, on 6 May 1937 at Lakehurst, following a transatlantic flight, broke the two golden rules of all disasters; by being one in the first place, and through being captured on film and narrated by a distraught reporter. The ship's high-profile demise spelt the end of airships altogether.

Every decade or so, the airship idea is briefly but unconvincingly revived. There have been plans for gigantic (and helium-filled) freight airships, and the Zeppelin company was recently reformed, but only with a view to building small blimps.

So by the Second World War, and greatly aided by preparations for it, the heavier-than-air machine had triumphed as the preferred form of flying.

This particular DC-3, seen here in the livery of Pennsylvania Central Airlines during the 1930s, remained in service until the late 1970s. Many are still flying today.

Two aeroplanes, examples of which are still in service today, illustrate the speed of airliner development as the world accelerated towards another war.

One is Britain's de Havilland Dragon of 1933 and its successor, the slightly improved Dragon Rapide of 1934. Its design, with its angular nose and tapering wings, appears to have been influenced more by Art Deco fashion than aerodynamic expediency, but in fact it flies rather well. It was the EasyJet of its day, and the origin of the European weekend away break.

But fly in a Dragon, and you will quickly realise that it still owes a great deal to the technology of the early ex-bomber airliners. Pull away the thin interior trim and you will discover metal tubes, wood and doped fabric. It is a biplane, braced with wire, with fixed undercarriage and powered by two small in-line engines.

A year later, America produced the Douglas DC-3, one of the world's most successful aeroplanes. The DC-3 was an all-metal monoplane, with retractable wheels, modern radial engines and efficient variable-pitch propellers. Some offered in-flight berths and a galley. In the early days of American air travel, the flight crew would carry train tickets to give to passengers if the flight was diverted or forced by weather to land short of its advertised destination. With the DC-3, travellers had a genuine alternative to the railroad. The American continent could be crossed in three hops in fifteen hours.

The designers of the DC-3 reckoned on building fifty to a hundred aircraft. Over 13,000 were eventually built and 400 remained in service in the late nineties. If any aeroplane can be said to have established the supremacy of air travel, it is the DC-3.

The war helped, since the DC-3 was also built as a military aircraft, designated the C-47. As such, it served as a paratroop transport, glider tug and general workhorse in all theatres of the world war. The Second World War, inevitably, spurred aeroplane development just as the first had and, just as importantly, provided a glut of highly trained pilots. By its conclusion the jet engine had been proven along with the pressurised cabin, a combination of which could be used to build an airliner that would fly faster and more economically at high altitude, while avoiding the turbulence that characterised low-level flight to boot. A pressurised fuselage meant the end of the huge windows and panoramic view of the Dragon Rapide, but in return passengers were far less likely to throw up en route.

In the immediate post-war years, as in the early twenties, small airlines struggled on with the DC-3 and yet more hastily converted bombers such as the Avro Lancastrian (an Avro Lancaster with a modified fuselage, much used during the Berlin Airlift) and the Vickers Viking (based on the wings and engines of the Wellington and employing its geodesic structure, which was the creation of Barnes Wallis, the R100 engineer). But by 1949 the world's first jet airliner, the de Havilland Comet, was ready.

The Comet was like no other civil aircraft before it, and not just because it had no propellers. Its top speed was almost 500 mph and it could fly at 40,000 feet. It entered service on the route between London

'IF ANY AEROPLANE CAN BE SAID TO HAVE ESTABLISHED THE SUPREMACY OF AIR TRAVEL, IT IS THE DC-3'

A de Havilland Comet coming together nicely in 1954. Sadly, it was to come apart rather quickly, too.

and Johannesburg, carrying 30,000 delighted passengers in its first year and generating orders for fifty aircraft.

But the jet age would have its disaster, and the first jet airliner would provide it. The cause of two mysterious crashes, both involving the catastrophic failure of the Comet airframe at high altitudes, was eventually traced to metal fatigue around, among other things, the corners of the aircraft's square window apertures. It was a harsh lesson in the structural laws governing a pressurised fuselage, and the reason why the windows on airliners today all feature radiused corners.

The Comet, after considerable modification, achieved another first in 1958, becoming the first jet airliner to enter transatlantic service. That year also saw the introduction, with Pan Am, of the legendary Boeing 707, the aircraft that kick-started the Seattle firm's domination of the world's airliner fleets. More commercially successful than the Comet, which inaugurated the jet age, it is the airliner that sounded the death knell of the post-war propliner.

Some of the world's most charismatic airliners disappeared with the coming of the jet, all relegated to history by the presence of their propellers.

'SOME OF THE
WORLD'S MOST
CHARISMATIC
AIRLINERS
DISAPPEARED
WITH THE COMING
OF THE JET AGE'

They included the Lockheed Constellation of 1943, the Douglas DC-6 of 1947, and the Boeing Stratocruiser (also from 1947, and derived from the B-29 Superfortress of Enola Gay fame). In their place came such classics as the Hawker-Siddeley Trident (1962), the Vickers VC10 (1962), and the Boeing 737 (1967). But the two greatest airliners of the twentieth century were yet to appear, and only one of them was to survive.

The Boeing 747, the jumbo jet, is a truly remarkable aeroplane, and not simply because it remains the world's biggest airliner nearly forty years after its maiden flight (with the right seating arrangement, it can carry over 500 passengers). No less impressive is the speed at which it was developed. In 1966, when airline travel was booming, Boeing gambled the entire value of the company on the 747. It entered service with Pan Am less than five years later. Work progressed at such an astonishing pace that the team responsible for the 747 became known as 'the Incredibles'.

There were many engineering and operational challenges in the 747. The double-decked fuselage was the first Boeing had employed since the Stratocruiser.[5] New engines were needed for such a monster, and were designed by Pratt and Witney. Ingenious flaps were needed to slow the 747's approach speed to one that would allow it to use existing runways, and the rakish sweep of its wings was necessary if it were to fit in existing hangars.

More mundanely, some new training methods were needed, among them a device known as Wadell's Wagon (after Boeing's chief test pilot, Jack Wadell). It consisted of a mocked-up 747 cockpit mounted on top of a truck and allowed prospective pilots to become familiar with the sensation of taxying an aircraft from a position thirty feet above the ground.

The first commercial flight of a 747 took place on 21 January 1970, in the same year as those of the largely forgotten Douglas DC-10 and Lockheed Tristar, on the Pan Am route from New York's JFK to London's Heathrow (and a rapturous reception). Boeing predicted sales of around 400 747s, after which it expected its mother ship to metamorphose in a cargo carrier. At the end of 2006, 1,380 747s had been built and a further 120 were on order.

The reason for Boeing's pessimism over the life expectancy of the 747 was the then widely held belief that supersonic travel was just over the horizon. Boeing itself was working on a Mach 2 swing-wing design called the SST. The Russian manufacturer Tupolev was working on its ill-fated Tu-144, and the year before the 747 entered service an Anglo-French consortium had successfully flown something called Concorde.[6]

Boeing eventually cancelled the SST, the Russian aircraft was blighted by a crash at the 1973 Paris Air Show and made only fifty-five scheduled passenger trips, so it was left to Concorde to blaze the trail to the supersonic future.

The magnitude of the threat from Concorde could be expressed in pure numbers. It would cruise at Mach 2.02, or around 1,330 mph, at which speed it would cross the Atlantic in half the time taken by a 747. It would fly at 60,000 feet, far above other aircraft, where it would enjoy dedicated air corridors. It was fast enough to outpace the diurnal spin of

Wadell's Wagon – pilots had to get used to taxiing an aeroplane from 30 feet off the ground.

the earth, which eventually led operator British Airways to boast that you could 'arrive before you leave'. With Concorde, it was claimed, no two cities on Earth would be much more than twelve hours apart.

It would also be very expensive. So expensive that no one aircraft manufacturer could possibly afford its development. It began as a concept by British Aircraft Corporation, but the British government of the time demanded that an international partner be found to share the cost. Only the French were interested, and signed a treaty with Britain for the development of the new aeroplane in 1962. Such was the excitement generated that this alone produced a hundred provisional orders from the leading airlines of the day.

The French prototype, Concorde 001, made its first flight on 2 March 1969, an event broadcast live on British television with commentary from the air show and technology veteran Raymond Baxter: 'She flies, she flies!' This simply increased the size of the prospective order book even further. Supersonic transport was a future no country wanted to be without. Already there were mutterings from some aviation commentators and politicians about the enormous cost of the project, but this was an era in which the gods of progress were to be served without concern for the burden of consumption.

But it was the issue of consumption that dealt the first blow to Concorde; more specifically, a Middle East crisis that saw a huge leap in crude oil prices. Supersonic aircraft, with their insatiable thirst for fuel, became about as fashionable as V8-engined American muscle cars. Furthermore, nations as far apart as India and the US were expressing concern over the noise generated by Concorde's four military-spec engines. By the time of the inaugural Concorde service from London and Paris to New York, an anticipated production run of 400 aircraft had been reduced to orders for just twelve, and those from the national airlines of the two countries that built it.

But none of this detracted from the undeniable allure of Concorde. Though an immensely complex aerodynamic proposition, in flight it exhibited the simplicity of a paper dart. Concorde trivia was of a sort that other aircraft couldn't deliver. When Concordes passed the magic Mach 2 mark, their cabin crews would take a discreet look at the passengers to see if anyone had exploited the uniqueness of the moment as an opportunity to propose. Concorde livery was always mainly white, because it helped prevent the aluminium skin from overheating at supersonic speeds. Even then, it expanded so much that the fuselage grew in length by a foot at Mach 2, and on its final supersonic trips in 2003 its flight engineers wedged their caps into the gap that opened up between their consoles and the adjacent bulkhead, so that when the airframe contracted their headgear would remain permanently lodged in the aircraft they had served.[7]

And to Concorde's transatlantic regulars – the film stars, media moguls and captains of industry – its cabin was like an exclusive club, and never mind that the seats were narrow, the overhead luggage space limited, or

James May's Twentieth Century: The Great Departure

the windows oppressively small. To the end of its days, Concorde was accepted as providing a greater sense of occasion than any other means of transport devised; perhaps than any other event. Even when it was thirty years old, a seat on Concorde placed you at the cutting edge, in the fastest commercial aircraft of all time.

It is perhaps ironic, then, that the fastest commercial aircraft of all time came to grief when it was going too slowly. The infamous Air France Concorde crash of 25 July 2000 was the result of a catastrophic chain of events that began with a small piece of debris on the runway at Paris Charles de Gaulle. This shredded a tyre, parts of which in turn punctured a wing fuel tank and started a fire. One engine on the port wing was shut down by the captain in response to the fire warning. Soon after, the second engine on the same wing failed. The combined problems of asymmetric thrust and drag from the undercarriage, which failed to retract, caused the airspeed to drop and the aircraft to sink, nose high, until it struck the ground. All 109 people on board, plus four on the ground, were killed.

The Boeing 747 'Jumbo Jet' takes to the air, 9 February 1969.

James May's Twentieth Century: The Great Departure

The crash, plus dwindling passenger numbers following the attack on the Twin Towers and an increasing maintenance burden, led to the retirement of Concorde in 2003. Inevitably, it was cited as the end of an era.

And it was, but it is worth remembering that the century in which powered heavier-than-air flight was born ended while commercial air travel was still supersonic; that the achievements of the Wrights, Blériot, Alcock and Brown and countless other pioneers, and the development of public air transport to the point were it is considered almost humdrum by many, all occurred within one lifetime.

A good measure of how quickly the aeroplane shrank the world might be this: it is perfectly possible that a passenger on Concorde, flying from London to New York at Mach 2, would have been able to remember reading in the paper about an earlier crossing of the Atlantic made by two men in something called a Vickers Vimy.

Goodbye.

THE PEOPLE'S POPULAR LIBERATION MOVEMENT

March 1936 – Margaret Allan on the Brooklands banking in a modified Bentley, practising for the following day's race meeting. I met her in 1992, when she was in her 80s and still a fairly fearsome driver.

Below:
Meanwhile, more demure ladies were encouraged to wear a fetching hat and drive a sensible car.

An hour in a Model T Ford is enough to make you wonder how this car lark ever caught on the way it did.

The Model T was designed and built 'for the great multitude', as Ford himself put it, and showed that cars might, after all, be accessible to ordinary people. It pioneered the moving production line, and put the planet's most forward-thinking people on wheels. So in a sense it was an integral part of America's pioneering attitude *per se*.

What it didn't pioneer, however, is the standardised system of car control we now take for granted, and which is as thoroughly cemented in the motor cortex of drivers as the act of walking. This can cause problems.

The Model T features three pedals, and to the right of the driver is a lever that could be mistaken for a simple handbrake. The right-hand pedal, which ought to be the throttle, is actually the brake. The middle pedal, which deep-rooted instinct tells you must actually be the brake, is a device for reversing the vehicle. The left-hand pedal, conditioning tells you, is the clutch; in fact it effects the shift between the car's two forward gears and neutral.

Or does it? It all depends on the position of that lever you thought was the handbrake. In the middle of its ratcheted quadrant, it gives first (pedal depressed) and neutral (pedal released). But if it is pushed fully forwards, that same action gives you a shift from first to second. Pull the lever fully back – and this is worth remembering, as it's the last-ditch get-out clause in that moment when all seems lost – to select neutral and apply the parking brake.

So where *is* the throttle? That's the small lever sprouting from the centre of the steering wheel, which is not to be confused with the other small lever sprouting from the centre of the steering wheel, which is used to retard and advance the ignition.

Practising around a car park for half an hour will convince you that you've cracked all this, and you will have done, at least with that creaking part of the brain given over to considered thought. But the road is more challenging and requires that the act of driving has passed into that area

'LOTS OF MODEL Ts RAN OVER THEIR OWNERS'

of subconscious action that some psychologists refer to as the 'body brain', and which faculty, in all but die-hard Model T enthusiasts, is totally beholden to the control layout of the Ford Focus.

There comes a horrible moment, as several everyday hazards present themselves simultaneously, when the new-found knowledge evaporates to be replaced with a blinding panic in which the feet fly to the 'clutch' and the 'brake' to give first and reverse simultaneously. But a built-in 'safety' feature seems to mean you get first. Meanwhile, you 'indicate right' with the throttle lever and accelerate towards whatever it was you wanted to avoid. At this point, obviously, it is best to heave back on the lever and bring everything to a halt. But you've forgotten that as well.

For the uninitiated, driving a Model T is really very hazardous. Almost as hazardous as starting the thing. It's hand cranked, like all early cars, and the thumb of the cranking hand must be kept tucked in next to the fingers lest a backfire reverses the engine's rotation and dislocates it. The ignition switch must be set to 'coil' and not 'magneto', otherwise you will crank until the end of time and achieve nothing.

A grand day out: in this case to Calhoun Bathing Beach, courtesy of the mass-produced motor car.

Most importantly, the throttle must be set just above idle and that lever must be in its rather indistinct central position, otherwise when the engine starts the car will simply move off and run over you. You wouldn't be the first to suffer this ignominy. Lots of Model Ts ran over their owners.

Despite this, there was a time in the 1920s when, it is estimated, half of all vehicles on American roads were Model Ts of some description (the Model T was available in numerous guises with a variety of car bodies featuring various seating arrangements and as a light commercial vehicle as well). Plenty of Model Ts survive, they can still be used legally on the roads, they can be refuelled at a regular petrol station and spares are still being produced. It was 'future proof', as the computer business would put it. That claim cannot be made of many other consumer artefacts dating from 1908.

Undoubtedly, the Model T was the most significant car of all time. (It was voted as such by an international jury in 1999.) This accolade was not so much because of what it was, but because it was the harbinger, and in many ways the instigator, of the most remarkable social phenomenon of the twentieth century.

The invention of the car is not an event that belongs to the twentieth century. At best, it belongs to the nineteenth, and in conceptual terms is certainly much older than that. But it was only in the second half of the twentieth century that car ownership began to look like something that could be presumed.

Although the Model T now seems terrifyingly arcane, it nevertheless established some basic automotive design parameters that, with the exception of a few experimental models, have never really changed. The car, unlike the aeroplane or the ship, was never turned into a significant instrument of war, so never benefited from the quantum technological leaps that war tends to force upon anything useful to its pursuit. In times of war, car factories are among the first industrial concerns to be commandeered for the production of other things.[8]

This woman may be only seconds from a horrible accident.

Overleaf:
Many Model Ts never went anywhere near what we would call a real road.

The electric car – not such a new idea as everyone thinks. This one is from 1902.

So the first Model T of 1908 exhibits a few basic features that we would regard as given: an internal combustion engine run on fossil fuel and using reciprocating pistons,[9] four wheels, a gearbox of some description, and control via pedals and a circular steering wheel.

Broadly speaking, the car has been like that ever since, but at the beginning of the century there was no indication that it would inevitably go that way.

Here are a few pub-quiz questions:

1 In 1899, in the US, which type of powerplant was most common in cars – steam, petrol, or electric?
2 In 1900, which country was the world's biggest producer of cars?
3 In 1902, what was the world's fastest car and how was it powered?

The answers are:

1 *Electric – there were 1,500 electric cars as against 936 powered by 'gas'.*
2 *France – ever keen to embrace a new idea.*
3 *The Gardner-Serpollet steam-driven Oeuf de Pâques, holder of the world land speed record at 75.06 mph. (Steam reclaimed the land speed record in 1906 when a Stanley Rocket Racer achieved 127.66 mph.)*

An even trickier question, and one unsuitable for pubs since there is no definitive answer, is: who actually came up with the idea of the car in the first place?

The correct answer could be the thirteenth-century Franciscan Friar Roger Bacon, who posited the idea of carriages moving without animals to pull them along. Or it could be the ubiquitous Leonardo da Vinci, who sketched a self-propelled vehicle. Or perhaps the Jesuit missionary Ferdinand Verbiest who, around 1670, developed a small steam-powered trolley.

Nicholas Joseph Cugnot certainly deserves a mention. In 1770 he built his Farrier à vapeur (steam dray), a self-propelled artillery carriage for the French army. This hefty three-wheeler was certainly a car in the loosest sense; more significantly, it was involved in the world's first recorded car crash, when it ran out of control at its heady top speed of 2 mph and demolished a wall. The Farrier survives to this day.

But the car as we know it – that is, a car that can be viewed as a direct ancestor of Ford's Model T – was the work of Karl Benz, whose Motorwagen was completed in 1885 and, more significantly, patented the following year. First, he had to design and build the engine that would dismiss the horses from their task at the front of the carriage, a job he had completed six years earlier but with some difficulty. On New Year's Eve 1879, his wife Bertha – clearly an incurable romantic – persuaded him to return once more to his workshop and have one more go at starting the contrary device. He records that his heart was pounding as he turned the crank.

'*The engine started to go put-put-put, and the music of the future sounded with regular rhythm. We both listened to it run for a full hour, fascinated, never tiring of the single tone of its song.*' It's a wonder the Benzes found the time or inspiration to produce five children.

In one sense Benz was right about the music of the future: by the end of the following century there were an estimated 500 million cars in the world. But his single tone was out of tune, because his engine worked on the two-stroke cycle, a type that has never found much favour in cars.[10] By the time Benz had built the Motorwagen he had adopted Nicolaus Otto's four-stroke design, and the engine mechanic's mantra of 'suck, squeeze, bang, blow' was unshakeably in place.[11]

Twelve years later Rudolf Diesel secured a patent for his compression-ignition engine. The fuel-efficiency benefits it offered would not be recognised by car designers for some time, but it, too, would become part of the music.

Benz claimed authorship of the car by the skin of his teeth. Gottlieb Daimler, Wilhelm Maybach and Siegfried Marcus were working on similar ideas at around the same time. Benz, though, proved his genius in the details (and his business acumen in applying for patents) and is credited with inventing, among other things, the spark plug, the clutch, the carburettor and the gearshift. Like the Wright brothers, he was yet another instinctive man-in-a-shed type of technician.

But it seems he was a little reticent about actually using his creation, so it fell to his wife Bertha to become the first motorist. In 1888, and unbeknown to her husband, she took the Motorwagen and, with her two young sons, drove from her home in Mannheim to Pforzheim, a distance of seventy-five miles.

James May's 20th Century: The People's Popular Liberation Movement

En route she persuaded a cobbler to re-shoe the primitive brake and cleared a fuel blockage with her hat pin. On its first outing of any significance, the car was alternately applauded and reviled, as it is to this day. But at its triumphant entry to Pforzheim, there was no doubt it had arrived.

So by the dawn of the twentieth century, the car, essentially as we know it, already existed. But it was still a long way from becoming a consumer durable. There were obvious barriers to widespread car ownership, including all the usual objections about frightening horses and what have you. It was also believed that car production would be limited by the availability of leather for seats and the supply of properly trained chauffeurs.

More significantly, the car was still something of a cottage industry product; something that was built rather than manufactured. All early cars were unique and assembled by skilled teams of fitters who fettled each part to a good fit. For that reason, a given design would often suffer from 'dimensional creep' – as its makers became more adept at its construction, they needed to remove less material to mate the parts together. So the finished article grew minutely.

Today, we accept that an off-the-shelf spare for a car engine will fit without further ado, but in the early days this was not so. And before Henry Ford could mass-produce his Model T, this whole business of parts interchangeability had to be sorted out.

It already had been in industries such as clock-making and small-arms manufacture, where carefully controlled tolerances ensured that, for example, the balance wheel of one clock would fit perfectly in another clock of the same design. The same ideas were championed in car-making by, most memorably, Henry Leland, an American engineer and machinist.

Leland demonstrated the benefits of parts interchangeability by stripping a number of his engines, mixing the parts up, and then reassembling them without needing to modify any of the components. They all ran happily. It meant that independent garages would be able to service and repair cars with stocks of standard parts. The company Leland worked for was called Cadillac which, curiously, had sprung from an earlier venture by one Henry Ford.

So Henry Ford did not truly invent mass production, since it was dependent on the principle of 'limits and fits', which was already well established. Neither was he responsible for the first series-produced car, which was Ransom E. Olds' Curved Dash model of 1901. (In its first year 425 were produced, a huge number by the standards of the time.) What Ford gave us was the moving assembly line, the most significant contribution to the building of cars in such number and at such a price that they might become commonplace.

And even that wasn't entirely his idea. It was suggested to him by William Klann, who had observed a 'disassembly line' in a Chicago slaughterhouse, where animals moved past stationary workers who were each charged with one butchering task. Ford reversed the process, allocating each worker a ruthlessly simple assembly operation to perform on cars that moved past them on a conveyor.

The effect was momentous. The first Model Ts, which were still assembled by teams of workers in the established way, were launched in 1908 at a cost of $825. By 1916, a Model T rolling off the new assembly line could be yours for $360. Between 1913 and 1914 the time taken to assemble a car dropped from twelve man hours to little more than ninety minutes.

Ford's view of his factory workers seems horribly unenlightened by today's standards: 'I could not possibly do the same thing day in and day out. But to other minds, perhaps I might say the majority of minds, repetitive operations hold no terrors. In fact to some kinds thought is absolutely appalling. To them the ideal job is one where the creative instinct need not be expressed.'

At the time, he was something of a saviour. His promise of a 'five-dollar day' caused a mass migration of unemployed labourers from the south to the Ford factory. He paid his workers in proportion to the output of the factory, creating a ready market for its product. When production ceased in 1927, 15,007,034 Model Ts had been built, a record that would stand until surpassed by the VW Beetle in 1972.

So the Volkswagen, though the first to be named as such, was not actually the first 'people's car'. The Model T was. It proved, barely twenty years after its invention, that the car had the potential to become as accessible as the refrigerator or radio would. But that wouldn't happen immediately, and neither would it happen suddenly.

The problem with any overview of the car's development is that it lacks real drama. Over the same period, aircraft were completely transformed with new materials, radical changes in design, new powerplants and a twenty-five-fold improvement in performance. Surgery progressed from little more than glorified butchery to the era of transplants, artificial organs, keyhole incisions and laser treatment. The field of consumer electronics was transformed several times, not least by the progression from valve to transistor to microchip. But the car simply lumbered on, improving hugely, certainly, but only by increments.

There was motorsport, which today can be seen as a glorified manufacturers' advertising campaign, but which was once a credible arena for proving the durability of cars you might actually buy from a showroom. Brooklands race circuit, Le Mans and the Gordon Bennett Trophy all thrilled crowds and advanced the cause of motoring, but the car itself still only progressed in short steps.

Consider this: the single most radical production car of the twentieth century was almost certainly the Citroën DS of 1955. Next to the dowdy offerings of its European neighbours, still hamstrung as they were by pre-war technology and a post-war austerity of thought, it was amazing . In France, its launch was considered the biggest news story since the death of Stalin (on the day of its launch at the Paris Motor Show, 12,000 orders were taken).

The Citroën pioneered some ideas that were still being touted as new by other manufacturers at the end of the century – headlights that swivelled with the front wheels, for example. It used a high-pressure hydraulic system for its brakes, transmission and self-levelling and adjustable suspension.

The launch of Citroën's remarkable DS, at the Paris Motor Show, 7 October 1955.

It had a semi-automatic clutchless gearbox and made great use of the new plastics; not just in its interior but also in its body construction. The styling was, and still is, fabulous in the real sense of the word (which is why it appears on the cover).

And yet … it was still driven by an internal combustion engine run on fossil fuel and using reciprocating pistons, it still had four wheels and a gearbox of some description, a circular steering wheel, and it was still controlled via pedals.[12] The contrast between the DS and Ford's Model T is nothing like so great as that between the Hawker Hunter and Blériot's cross-Channel monoplane, even though the developmental timespan is roughly the same (actually, slightly shorter in the case of the aeroplanes. The Hunter first flew in 1951).

In fact the story of the car is far more interesting when it is viewed as a social rather than a technical phenomenon. The miracle of the car has always been its ever-increasing accessibility, the liberty it affords ordinary people, and the impact it has had on society as a result, not in what's going on under the bonnet.

'PERHAPS MOST BEETLE OWNERS NEVER KNEW THAT THE WORLD'S MOST INFAMOUS DICTATOR HAD MADE HIS FINAL JOURNEY IN ONE'

So to car buffs, significant vehicles might be the Rolls-Royce Silver Ghost of 1906, which proved the long-distance touring potential of the idea; the Bugatti Royale of 1929, for being the most sumptuous car ever built; the Mercedes 300SL Gullwing of 1957, for being the first true supercar; and the McLaren F1 of 1993 for being the fastest road car of the century.

But to social historians, mundane and everyday cars are more revealing. The quest for volume and affordability has taxed great automotive minds far more vigorously than the goals of performance and exclusivity, and has often yielded more significant results. This is especially true of those cars that carried the mantle of 'people's car', as established by the Model T.

Britain's best contender for the title of original people's car is the Austin 7, which was produced in a bewildering array of variants and for seventeen years, beginning in 1922. It has been called Britain's Model T and for good reason, for although it was produced in relatively tiny numbers (a total of 290,000) it did a similar job in persuading people that they could own a proper car and not just one of the hideous cyclecars that proliferated in the 1920s.[13] The Austin 7, and its pricing, are credited with demolishing the British cyclecar industry, along with countless now forgotten small-volume manufacturers of the era.

Italy has produced several people's cars, but most famously the diminutive Fiat Cinquecento 500 of 1957. Italy, which in the war enjoyed the distinction of having its industrial base flattened by both sides, struggled to mobilise its people in peacetime. The Vespa scooter was the first solution; Fiat's range of small cars was the next. The 500 was in effect a car built with a scooter mentality – painfully simple, yet roomy enough for four (just), cheap to run and easy to maintain. It was so popular that in remote areas corner shops often stocked the essential spares. It remained in production until 1975.

Two years after the launch of the Fiat, Britain would produce the Mini (known originally as the Morris Mini-Minor and, auspiciously, the Austin A30 or 'New' Austin 7), largely in response to fuel shortages brought on by the

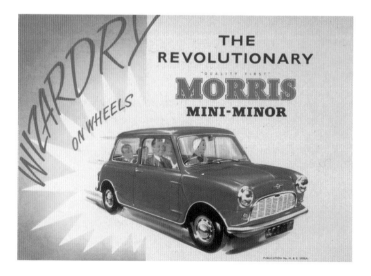

Suez Crisis. The packaging advantages of its transverse-engined and front-wheel-drive layout would change the future direction of virtually all small cars and quite a few big ones, too. It has been called the 'classless car' to the point where the appellation has become hackneyed but it was exactly that (and yes, Twiggy had one). It stayed in production until 2000, and the car that was described by its designer as being 'for the housewife' quickly came to be regarded as something of a driving enthusiast's machine.

France, with its overt socialist leanings, was especially good at people's cars; not just cars cheap enough to be owned by the masses, but cars designed to suit the rather medieval lifestyle of its huge rural population. Most famous is the Citroën 2CV, of 1948–90, whose simple air-cooled engine, bolt-on panels, narrow wheels and long-travel soft springing were all incorporated to conquer the nation's rutted and unsealed post-war country roads. Part of the design brief was that the 2CV had to be able to carry a basket of eggs across a ploughed field without breaking any. (The originator of the design, Pierre-Jules Boulanger, later raised the height of the roof so he could drive without removing his hat.)

More numerous, more important but less celebrated was the Renault 4 of 1961, considered the most successful car in French automotive history with production exceeding 8 million. It remains, despite going out of production in 1993, the quintessential French town and country car, as likely to be seen transporting a live pig as a holiday-making family.

Curiously, Citroën's 2CV was taken to the prototype stage before the Second World War, but was then hidden during the occupation for fear that it would fall into the hands of the Nazis and be put to militaristic use. In fact the Nazis had their own people's car. One day, it would even come to be named as such – the Volkswagen. The Beetle was to become the best-selling car of all time.

But to begin with, it had a different name. It was conceived as part of Germany's 1930s *Kraft durch Freude* (literally, 'Strength through Joy') movement, a programme designed to make leisure activities affordable for the people. It would be called the KdF-Wagen, and would be built in a new factory, complete with workers' housing, called KdF-Stadt.

The proposal was Hitler's. He wanted a cheap and simple car that could travel at a sustained 100 kmh and carry two adults and three children, and which could be bought by virtually anyone through a simple state-controlled savings plan. He might have added that it would be an ideal way for Germans to drive up the arrow-straight Autobahn he envisaged running from Berlin to Moscow.

Ferdinand Porsche produced a car for him, a small streamlined car with an air-cooled engine. Or did he? It is now accepted that the design was stolen from the brilliant Czech engineer Hans Ledwinka of Tatra, who had already produced a remarkably similar car, called the T97, in 1931.[14]

The story of the Beetle is as scandalous as it is fascinating. Backed by the Third Reich, KdF-Stadt began producing its new car in 1939, but few of the German people who entered into the savings scheme ever gained strength through the joy of driving one. A handful were built for the Nazi

gowns by Ceil Chapman

RITZ-CARLTON

It would be difficult to recount *all* the wonderful things that a Cadillac car indicates about its owner. But it is readily apparent, we think, that it now speaks more eloquently than ever of his good taste and judgment. Cadillac's new beauty, for instance, is graceful and inspiring as never before. This new Cadillac refinement is equally evident in the car's interiors—in the rare quality of its fabrics, leathers and appointments . . . and in the care and skill of its Fleetwood tailoring and craftsmanship. And then, of course, there is the car's marvelous new performance and handling ease. We suggest you visit your Cadillac dealer soon. You'll quickly see why the new "car of cars" has been accorded the most brilliant reception in Cadillac history.

CADILLAC MOTOR CAR DIVISION • GENERAL MOTORS CORPORATION
EVERY WINDOW OF EVERY CADILLAC IS SAFETY PLATE GLASS

The world's most eloquent possession... *Cadillac*

elite and senior officers, but for the most part the new factory was given over to producing military vehicles based on the Beetle chassis. These included the Kubelwagen (beloved of Rommel and his Afrika Corps. The air-cooled engine performed well in the desert) and the amphibious Schwimmwagen. Eastern European slave labour and PoWs were drafted into the workforce.

It was only after the war that the KdF-Wagen was produced in number. The ruined factory – since it had also produced munitions, it had been heavily bombed by the Allies – was offered as a reparation to both Ford in the US and the Rootes Group in Britain, but both organisations considered it worthless. 'What we're being offered here isn't worth a damn,' declared Henry Ford himself. (He is later said to have called the Beetle 'a little shit box' when it became a best-seller in the US.) It fell to a British army officer named Ivan Hirst to rescue the tooling for the car and, with the help of some of the factory's original staff, put it back into production. The town was renamed Wolfsburg, and the car 'Volkswagen'.

'This vehicle,' Lord Rootes told Hirst, 'is quite unattractive to the average buyer. It is too ugly and too noisy. If you think you're going to build cars in this place, young man, you're a bloody fool.' But build them he did, encouraged by an order for 20,000 from the occupying forces. They admired the Volkswagen for the same reasons the German military had: robustness, ease of maintenance, air-cooled reliability.

So the car that came to be known colloquially, and then officially, as the Beetle became the people's car of the original plot, if not for quite the same group of people. It was built under licence in many parts of the world and remains the best-selling car of all time. The final irony of the Beetle story is that a car that started life as an instrument of the Reich quickly became something of an icon to the hippy and peacenik movements. Perhaps most Beetle owners never knew that the world's most infamous dictator had made his final journey in one.

By the fifties, America and Europe had adopted, largely through circumstance, slightly different attitudes to the car. America, a land of plenty, was still a relatively new nation enjoying the delirium of rapid development. Its people laid the road before them and, reasonably enough, made it straight. There was an excess of production capacity following the war effort, new wealth, optimism, a desire for modernity and a sense that owning a car might just be a God-given right. America was big, the distances were big, so the cars were big and indulged their occupants. Their styling was inspired, like the door knobs and light fittings of the motels where their owners stopped, by the design language of the new jet age (America has never made a truly successful small car. European cars have found great favour over there, but the opposite has rarely been true).

In devastated Europe, the offerings were more prosaic. Worn-out factories, penury, the lingering spectre of fuel shortages and rationing, and the need simply to rebuild confirmed the car as still something of a privilege. Europe dragged its roads in its wake, like the evolutions of wandering, ancient pathways that many of them were. Europe's offerings were perhaps more inspired as a result. Small, simple cars with small but willing engines made the most of the situation, the roads, and most people's budgets.

By the mid-fifties inspiration was clearly re-emerging from the fug of post-war thrift. The Citroën DS appeared like a wheeled spaceship from a more modernist galaxy, ahead of its time but a sign nevertheless that things were looking up. European cars of the fifties finally began to shake off their fuddy-duddy demeanour.

By the sixties, the transformation was complete. Numerous cars could be held up to illustrate the final liberation of the European motoring family, but none celebrated it quite like the Mk1 Ford Cortina of 1962. The Cortina was not a very radical piece of engineering. The engine and suspension were unremarkable; it did not advance the state of automotive technology in any notable way.

Prior to the Cortina, British cars bore oppressive establishment names such as Westminster, Oxford, Cambridge, Prefect and Consul, and they were inevitably finished in sober hues. But here was a car named after a sunny and exotic Italian resort, plenty big enough for a family, reasonably fast, available in cheerful colour schemes and sprinkled with contemporary details such as vinyl upholstery and rear light bezels that looked like ban-the-bomb symbols. Significantly, it was a car suitable for the new motorways; roads laid, like America's, in anticipation of the car, not patched up in its wake. Cortina-owning families might take a day out to visit one of the new and futuristic service stations.

Just as the French philosopher Roland Barthes had eulogised the beauty and innovation of Citroën's DS, British poet John Betjeman saw fit to record the social significance of the new Ford:

> I am a young executive. No cuffs than mine are cleaner;
> I have a Slimline brief-case and I use the firm's Cortina.

Many periods in motoring history have been declared its 'golden age', but the epithet seems to belong best to the sixties and seventies, a time when, in Europe at least, car ownership was stripped of its class connotations and proliferated. The car became a relatively modest aspiration instead of a restrained badge of professional or middle-class office. It became as much fashionable as functional. It can be argued that, in real terms and set against improving standards of living, cars became cheaper. They also became more reliable, easier to buy and less burdensome to maintain. In short, the car had become more accessible. Henry Ford had made his point.

To the end of the century, the ownership issue dominated the development of the car. A progressively less misty-eyed consumer demanded better durability, better comfort and higher levels of equipment, spurred greatly by the examples set by the new car-building giant that had risen in the east, Japan. Attributes such as occupant safety and fuel consumption became as important as performance and styling in the ruthless arena of car marketing.

A Ford Cortina MkI, with its famous 'Ban the Bomb' tail light design.

Greater efficiency of the machine was achieved largely through greater efficiency in the manufacturing process – Ford's moving assembly line gradually gave way to a new doctrine of just-in-time manufacturing and cell-based production, in which the workforce that Ford had dismissed as incapable of thought became the arbiters of quality control. Even the financing of a new car became an enormous industry in its own right.

But what of the car itself, the car as an expression of humankind's ingenuity? With a few exceptions, it was still driven by an internal combustion engine run on fossil fuel and using reciprocating pistons; it still had four wheels and a gearbox of some description, and it was still controlled via pedals and a circular steering wheel.

But for how much longer? The car has never been entirely welcome. In its early days, people threw stones at cars. A hundred years later, they were hurling political brickbats, lambasting it for its slavish dependence on a finite fuel reserve and the damage caused in consuming it.

By the year 2000, there was compelling evidence to suggest that the car of the future might run on electricity generated by an on-board power station in the form of a hydrogen fuel cell, and might be steered and managed entirely – even autonomously – by computer.

The car, finally, might be poised to make its great leap forward.

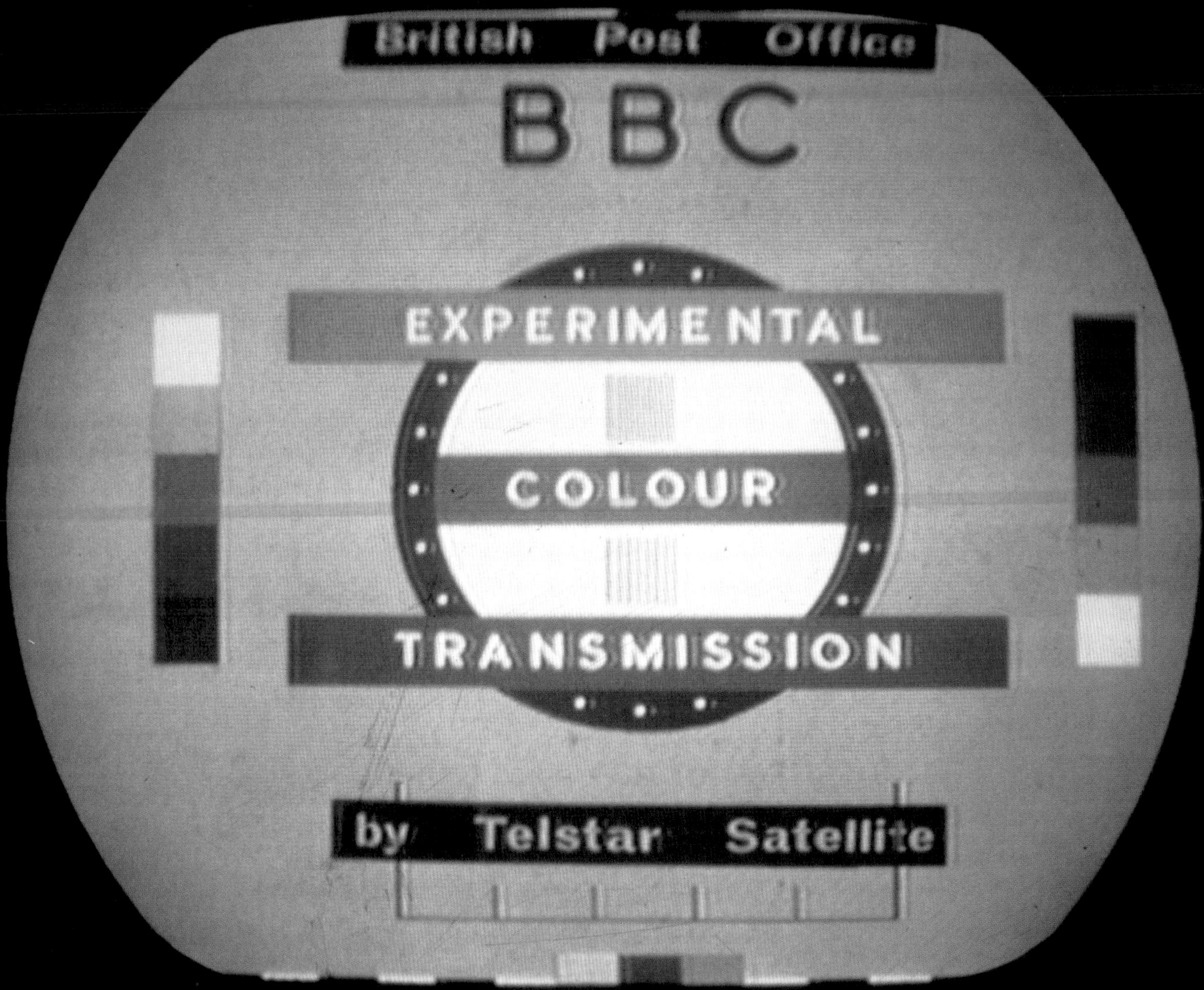
British Post Office
BBC
EXPERIMENTAL
COLOUR
TRANSMISSION
by Telstar Satellite

THE THINKING INSIDE THE BOX

Conceptually, the idea of television existed way back in the thirteenth century, when the idea of seeing an image of a distant event in your own front room was being investigated by a young Italian girl called Clare.

She was born into the aristocracy, but turned to God after hearing St Francis of Assisi preach about animal welfare. From then on she led a devoutly religious life, even contemplating martyrdom at one point. Instead, she lived to frail old age, but in doing so encountered a problem: she was too weak to attend mass at church and was confined to her humble cell. Fortunately, an image of the service would miraculously appear on the wall like an early *Songs of Praise*, but without Thora Hird.

Clare died in 1253, and was canonised two years later. As reward for her devotion she was given the heavenly brief of keeping a watchful eye over needleworkers, goldsmiths and, when the opportunity came along a few centuries later, she broadened her portfolio to become the patron saint of television.

The irony is that, compared with the first true (and temporal) television system, St Clare's vision was much closer to the phenomenon we know today; that is rather miraculous. Early television was a clunky and oddly mechanical affair, and its inventor should be credited more for coming up with the basic principles than for conceiving a viable system of transmitting pictures.

And who was the inventor? Like so many of the twentieth century's great innovations, television turns out to be the culmination of a lot of different people's work, all conveniently packaged in one remarkable box. In Tokyo people will tell you it was the work of Takayanagi; if you're in Moscow then it was Zworykin or Rosing; in Paris, it was certain to be Belin; in Germany it was Nipkow or Karolus. In New York they are not so sure; it was either Jenkins or Farnsworth. But here in Great Britain the answer you will hear most often is John Logie Baird.

The more you discover about John Logie Baird the more you have to like him. He had some hair-brained ideas but they usually had a touch of genius about them, too. He came up with a rust-proof razor, cleverly made of glass, but it broke all too easily. How about pneumatic shoes? He even tried to create diamonds by passing an electric current through graphite, only to short out Glasgow's power supply. On the other hand his thermal sock was said to be something of a success. (He usually worked in an attic, but still qualifies as a man in a shed. The difference is largely academic.)

Soon this born inventor set his sights on what would come to be known as television, or at least the basic principles of it – scanning an image, transmitting the information and reproducing it on a screen remotely.

Baird had been experimenting with something called the Nipkow Disc, a German invention of the late nineteenth century. It featured an arrangement of punched holes forming a spiral starting from its hub, so when it was spun at high speed it was possible to see 'through' it. But, significantly, whatever the observer saw through the disc was constantly dissected into a series of slices (one problem with the disc-based system however is that these slices will always be slightly curved rather than straight).

At the heart of Baird's transmitting device was one of these spinning discs. To pick up the image seen through the holes he fitted sensors that responded to light. These created an electrical signal, the strength of which was proportional to the amount of light received, which could be transmitted to a receiver. In the receiver a light, modulated in accordance with the signal, was shone through a similar Nipkow Disc on to a viewing screen. Providing that the discs in the transmitter and receiver spun at the same

John Logie Baird in his shed/attic, hard at work on the 'Televisor'.

Below right:
Stooky Bill, the first 'face' of television.

A real human face, as it appeared on the televisor.

speeds, the picture would appear. This system is known as 'mechanical television' for obvious reasons, and is a truly analogue business. But then, Baird inhabited a truly analogue world.

Baird was convinced he could make mechanical television work. To improve the picture he made his scanning discs bigger and bigger, up to about eight feet in diameter. To improve the intensity of the light passing through the holes, he fitted a glass lens in each one, but even then the subject had to be lit very brightly indeed. With the whole assembly spinning at 750 rpm, it was important to stay away from the trajectory of any lens that worked loose, as they sometimes did, taking huge chunks out of his laboratory wall. The device was large, heavy and noisy and, with only thirty to fifty holes in the scanning disc, the picture quality was never going to be great.

Undaunted, Baird worked away in his attic laboratory above a restaurant in London's Soho and by February 1924 he felt ready to demonstrate his charmingly named 'televisor' to the *Radio Times*. Using his fingers to make shadow puppets, he showed that it was at least possible to transmit crude silhouettes.

Baird kept on improving his system. As the lighting at the transmitter end was extremely hot, he often used as his subject the head of a ventriloquist's dummy called Stooky Bill (Stooky's head survives in the National Media Museum in Bradford). Soon Baird could transmit a thirty-line vertically scanned image at a rate of five pictures a second. To put this in context, today's televisions display 625 scanned lines at twenty-five pictures a second.[15] The problem Baird faced, as ever, was one of bandwidth. His first 'transmission', between London and Glasgow in 1927, was made using the telephone network, with its obvious limitations. Thirty lines, he calculated, was just enough to render a recognisable face.

Baird persuaded a young man, twenty-year-old William Taynton, to sit in for him to see what a real face would look like on his screen, and thus Taynton became the first person whose soul was stolen by television. What's more, this wasn't just a stark black and white silhouette, there were even shades of grey in the picture. Video was still a long way from killing the radio star, but it was a start.

Although there were not yet any broadcasters, Baird was confident enough to give a public show of his invention at Selfridges department store in London in 1925. Here shoppers were encouraged to place an order for a televisor of their own. Baird gave his guarantee that the televisor sets would be delivered as soon as a broadcast service became available. There would be time enough in the future for people to complain that there was 'nothing on the television'. A year later, he demonstrated his television system to the Royal Institution. 'Well, what's the good of it when you've got it?' asked one member of the audience. 'What useful purpose will it serve?' Some people are still waiting for an answer.

Further improvements followed. Incredibly, Baird managed to transmit a colour image as early as 1928 (some say it was 1927) by using three spinning discs, each with a different colour filter. If that

'THERE WAS A DESCRIPTION BY A BUS DRIVER, MR L.A. STOCK, OF HOW HE BUILT A MODEL OF SIR FRANCIS DRAKE'S *GOLDEN HIND* FROM MATCHSTICKS'

wasn't enough, he then had a go at 3D stereoscopic TV. He even made the first transatlantic television transmission, from London to New York.

Around the same time the BBC began its first rather tentative TV broadcasts and, given Baird's astonishing progress, it is perhaps surprising that they adopted his thirty-line televisor system. At the time nobody was sure how many people were watching. After all, they could broadcast to only a very tiny audience of brave early adopters. Some were electronics enthusiasts who had built their own televisor sets, and others were, presumably, just rich; the sort of people who would later want to be the first to own one of the new Betamax video players.

After a few years of these experimental transmissions, it was felt that television should be taken seriously and so, in true British style, a committee was set up to do just that. The first item on its agenda was which television system to back, since Baird's version – which had been greatly improved and now offered a 240-line picture – was rivalled by a different system from EMI Marconi, which boasted a stunning 405 lines. EMI Marconi's television didn't employ any spinning wheels, but instead exploited another German invention from the late nineteenth century, the cathode ray tube. This was a tube of glass with an electrode at one end and a fluorescent screen at the other; it looked rather like a massive light bulb. When the electrode was heated it sent out a stream of electrons that bombarded the fluorescent screen, creating a spot of light. By controlling the strength of the electron stream and directing it back and forth across the screen, it was possible to conjure up a picture.

It was proposed that both types of television should be tried out. Programmes would be transmitted by the BBC using alternate systems, Baird one week, and EMI Marconi the next, until it was determined which was the best. This might have been very fair and square, but it meant that manufacturers had to create television sets capable of receiving both types of transmission. To make their commitment to this new medium absolutely clear, the BBC decided it would move out of the temporary home in the basement of Broadcasting House, where the early trials had taken place, to a new home with a commanding view right across London, in Alexandra Palace.

So Baird's spinning wheel squared up against EMI Marconi's tube and on 2 November 1936 the BBC began the world's first regular television service, transmitting to an estimated one hundred sets around London, as then the signals could only reach forty miles. The service went out two hours every day, except Sundays. At the time not everyone in the BBC was in favour of this new upstart broadcasting medium. Even Sir John Reith, the Director General, did not appear to be fully behind his new channel, recording in his diary: 'To Alexandra Palace for the television opening. I have declined to be televised or take part ...'

Perhaps the reason for Sir John's caution could be found in the early schedules. Programmes showcased acts like the dancing comedians, Buck and Bubbles, or the Chinese jugglers, the Lai Founs. And there was a description by a bus driver, Mr L.A. Stock, of how he built a model of Sir Francis Drake's famous ship, the *Golden Hind,* from matchsticks. It's a wonder TV ever caught on.

With the two rival television systems operating side by side, the inherent weaknesses of John Logie Baird's mechanical system became ever more apparent. One notable drawback, at least for fledgling television stars, was the need to wear very bright blue and yellow face paint in the interests of picture contrast. Needless to say this went down none too well, especially amongst those presenters who, like so many of their successors today, forgot to take their make-up off before stepping out into the real world. In less than one year the Baird system was rejected and the EMI Marconi system, the basic principles of which survive to this day, triumphed.

Having seen his television system beaten so soundly, Baird might have felt inclined to give up. Instead, he kept on inventing. In 1944 he demonstrated, for the first time, a fully electronic colour television display, offering 600 lines. During the war he persuaded the British government that they should adopt his new 1,000-line 'telechrome' electronic colour system as the broadcast standard for the post-war years, assuming they won. This was very far-sighted and close to the quality of the very latest high-definition televisions around today, which offer 1,080 lines. But the Brits were not much in the mood for newfangled inventions in the austerity of the late 1940s, so the black and white, 405-line standard remained firmly in place until the 1960s.

Perhaps most astonishing of all, Baird had even invented the video recorder as far back as 1927. He didn't use video tape, as that would not be around for a few decades to come, instead recording his televisor pictures on a 78 rpm gramophone disc. This was extraordinary when you remember that the first domestic VCRs were not available until the 1970s. Baird managed to record only his thirty-line signal, so the picture quality

1938: if television hadn't been invented, people might by now have forgotten all about Neville Chamberlain's 'peace in our time' speech.

Overleaf:
A BBC cameraman can't believe his luck as he zooms in on the Windmill Girls, Alexandra Palace, 1946.

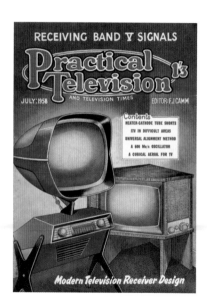

In the olden days, you had to
make your own entertainment.

was not great. A more significant shortcoming was that, although Baird had worked out how to make a recording, he had not been able to create a playback machine. But such was his faith in emerging technology that in the late 1920s he actually marketed 'video recorders' that could not replay the discs, confident that in time the appropriate machine would become available. These did not really catch on. The discs would remain unplayed for decades, but in the 1980s a group of scientists developed a way to decode the recordings. They revealed an eerie image of a tortured face – Stooky Bill, returning from beyond his owner's grave to confirm that his master had made the world's first video recording.

Once the BBC had only the EMI Marconi system to work with they began to get into some serious programming. In 1937 the coronation of George VI was transmitted to an estimated 9,000 sets. Then the Wembley Cup Final, between Preston and Huddersfield Town, was shown in 1938, as was the Oxford and Cambridge Boat Race. The cameras were also present for Neville Chamberlain's return from Munich and his infamous 'Peace in our time' speech. Sadly, Chamberlain could not have been more wrong; a year later Britain was at war with Germany, and one of the first casualties of the conflict was television. The nation's 20,000 sets went blank.

So television had shown that it had the potential to shrink the world but, in Britain at least, it was going to have to wait until the war was won. Once that little difficulty was dealt with and the Nazis thoroughly defeated television came into its stride. The BBC recognised, long before anyone else, that the key to a ratings success was some reality TV action from a big house, in this case Windsor. Britain clamoured to watch the 1953 coronation of Queen Elizabeth II, and in just one month over 2 million sets were sold. The television set was well on its way to replacing the radio as the focal point of the household and not everyone was happy about it. Some saw it as the Antichrist. And Mike Haywood, a poet from South Yorkshire, wrote:

> Shadows have been taken from lovers
> By the new lights.
> No wonder people marry
> So young nowadays
> And save to buy television's darkness.
> Is it a plot?

Meanwhile, in the US, television technology had leapt ahead, not least because it had not been interrupted by the war. In 1956 Ampex was ready to unveil a device that would at first change the way programmes were made and ultimately how they were watched – the video tape recorder. Fortunately, 3M announced the first magnetic tape to go with it.

The device was massive – about the size of a wardrobe – and used huge spinning tape reels, so was still a far cry from the domestic machines that would become popular in the 1970s. But it was a boon to broadcasters, and not only because it would liberate TV stations from the terror of live broadcasting (although it certainly did that). The problem was that the US is a very big country with different time zones, so in order to hit peak-time audiences in New York, Chicago and Los Angeles popular shows would need to be transmitted three times a day – a hideous tyranny when live performance was the only option. Video tape meant one recording session and broadcast at will.

Unfortunately, the machine and its tape remained extremely expensive resources until the mid-1970s and so, hard though it is to believe now, in Britain many comedies and dramas were actually transmitted live throughout the sixties. In big TV dramas of the day, such as the police series *Z Cars*, the detective often broke down the door of a bed-sit flat in the hope of finally nailing his villain, only to discover a rather embarrassed cameraman lurking on the other side.

But video recording did mean that British viewers could now watch selected programmes from all over the world, whether it was *I Love Lucy* from the US (which helped launch ITV in 1955) or the latest on the Suez Crisis. For the vast majority of people it was their first and only contact with life outside the British Isles.

However, it was not until the 1960s that television revealed the true extent of its ability to shrink the world. In July 1962, at the Goonhilly Ground Station in Cornwall, a group of pioneering engineers waited with bated breath in front of a small black and white TV monitor. The location was already deeply significant in the history of global communications: the first transatlantic telephone cable had terminated in Cornwall, and it was from Cornwall that Marconi had made his pioneering broadcast to Newfoundland.

And then it came, the image of a distorted face, redolent of Baird's Stooky Bill. The signal that produced it had been bounced from a newly launched satellite called Telstar (Telstar even had its own theme tune, recorded by the Tornados, and became a number one hit in the US and Britain). As ever, Telstar was the victim of bandwidth limitations, in that it

Below:
'That's a man's face! There it is!' said Raymond Baxter, perceptively, of the world's first satellite television picture.

Overleaf:
1967 and what appears to be an early sex education TV programme leaves the two youngsters looking confused.

could carry only one TV signal or sixty phone calls, but it did demonstrate that live events could be seen on television anywhere on the planet, given sufficient satellite coverage.[16]

One barrier to constant coverage was Telstar's elliptical orbit, which meant it could bridge the Atlantic divide for only around twenty minutes out of every two and a half hours. It was calculated that fifty-five satellites in a mixture of polar and equatorial orbits could provide uninterrupted communication, but at massive expense.

A better solution was a geo-stationary satellite, locked in one position above the earth and available twenty-four hours a day, even if that meant positioning it at an altitude of over 20,000 miles, instead of Telstar's lowly orbit. The world had to wait another three years before Early Bird, the first truly commercial satellite, was launched. It lasted for more than four years and, fittingly, its final task was to transmit the pictures of the first moon landing in 1969. Nations would not only speak unto nations, they would

keep an eye on each other's TV shows as well, even when they were coming from another planet.

At the time of writing, over 900 satellites are positioned above the earth, some of them secret military devices engaged in covert activity, but many of them happily beaming live sporting and cultural events from one side of the globe to another and confirming, like no other piece of technology, that there really is such a thing as the global village. At the end of the twentieth century there were estimated to be 240 million TV sets in the US alone and some programmes on China Central Television are seen by over 600 million viewers. The BBC has estimated that in Britain there will be 74 million TV sets by 2020, which is more televisions than viewers. Even the nomadic tribes of Outer Mongolia, who can pack up their entire belongings on to a few yaks, have satellite dishes next to their gers, the big white tents they call home.

At the beginning of the twentieth century natural historians were fond of pointing out that the entire population of the world would fit in a wooden crate one mile cubed. Now it seems that the planet itself might be crammed into a small box in the corner. Along with some adverts for soap powder and a panel game or two.

A NICE BIG BOX OF DIGITAL DELIGHTS

'THERE HAS
NEVER BEEN
A COMPUTER
OF ANY TRUE
AESTHETIC MERIT'

In the first half of the twentieth century the people of the world were granted fabulously visual manifestations of technological progress: the motor car, flight, television. But those of us who entered the century well into its second half had to be amazed by more abstract things – new materials, lasers, videotape and, most significantly, computers.

There is no doubt that the computer, in its broadest sense, has made a greater impact on our daily lives than any other branch of science and technology, not least because it plays an instrumental role in almost everything we do. Computers monitor our condition before birth, organise our lives in the tiniest detail, and might well be used to confirm our deaths, which will almost certainly be recorded on one somewhere.

What follows is a purely personal view. I can't stand the things, especially in their domestic form. I use one constantly, this book has been brought to you with the use of several of them, and they are doubtless marvels, the workings of which are almost impossible to comprehend. And yet they are strangely *unlovely*.

I hate the cussed logic of computers, that digital pedantry that leaves them soulless in a way a proper machine can never be. Computers are marketed as lifestyle enhancers and creative forces, when most of us know that they are really just tools and won't write this book any more than a hammer can make a horseshoe. Despite what people love to write about the Apple Macintosh, there has never been a computer of any true aesthetic merit. They also make an annoying ticking noise, fill our lives (or, more specifically, the region behind our desks) with a hideous jumble of wires, bombard us with jingles and smell funny.

This is not an entirely irrational phobia either. It is fairly easy to argue, as people have, that the home computer and its related devices have enslaved us and our wallets to a business that thrives on continued obsolescence. The increasing ease with which a computer chip can be incorporated into everyday devices has undoubtedly added unnecessary complications in some

German troops with an early laptop, aka a teletype machine, looking up the weather in Poland.

Wrens at Bletchley Park with the Colossus MKII computer, so called because of its compact dimensions.

areas of our lives, and while the computer's ability to process data should have saved so much labour that we can all afford the time to become classical scholars, it seems that far too many people spend their whole lives ministering to the device that was supposed to liberate them.

Still, as someone once said, love 'em or hate 'em, they're here to stay. My generation has been defined by the silicon chip, and the only consolation is that the whole business started with lovely cogs and levers.

The grandfather of the modern computer was a reassuringly nineteenth-century figure, Charles Babbage. In the early 1800s there was an escalating requirement in every walk of commerce, industry and science, to make calculations: in banking, insurance, bridge-building, surveying and astronomy, the need to crunch numbers was reaching something of a crunch point itself.

The only data-processing tool on hand were the much loathed logarithm tables.[17] Many businesses employed teams of arithmeticians, known literally as 'calculators', whose job was to perform complex maths homework for a living. Often many people would work together on one big problem, combining their finished work to come up with the final figures. Not only was this incredibly laborious, it also led to many errors.[18]

Babbage, still a twenty-one-year-old student at Cambridge, thought it would make sense to entrust this work to a machine of some sort and in 1821 came up with his Difference Engine. It had little in common with a modern computer, relying on gears and axles and looking rather like a clock with a great many 'complications' (as horologists would describe extra features such as a moonphase calendar). Babbage made a small prototype, but building the real thing was going to be a tough job. It required over 25,000 precision parts and, in any case, before work had progressed very far Babbage had come up with an improved design, the Analytical Engine. It employed much of the thinking seen in a modern PC, such as a memory and a central processor. Babbage even devised a mechanical printer to record results, thus denying errant humans a further opportunity to introduce errors to the business of calculating.

Sadly, none of Babbage's 'engines' was built. While the engineering of his day was first rate for bridges, steam engines and grandfather clocks, it was not yet ready to deal with the requirements of a computer, even a mechanical one driven by steam. But in 1985 the London Science Museum took it upon themselves to realise one of Babbage's visions, using the original designs, to celebrate the two hundredth anniversary of his birth. The design they chose was the Difference Engine No. 2; no small undertaking, as the calculating section alone weighed 2.6 tons, the whole thing was seven feet high and it relied upon 4,000 separate parts.

Remarkably, or perhaps predictably, it worked, digesting complex equations and producing correct answers (checked with a modern computer) up to thirty-one digits long. By preserving his design for posterity, Babbage had said 'I told you so' from beyond the grave. In fact, it was already well accepted that he had been right all along. The so-called pioneers of the modern electronic computer soon realised that Babbage had anticipated every essential feature of their work. Even if he never actually built any computers, he certainly framed the philosophy behind all of them.

Every generation that followed him continued the quest to free humanity from the drudgery of double maths as a way of life. But what neither Babbage nor his successors ever imagined was that, one day, long and complex arithmetic was the last use most people would have for a computer (although, in truth, that is all a computer really does, isn't it?). But this change was still well over a century away.

In the twentieth century computing had a new Babbage in the form of Konrad Zuse, a twenty-four-year-old engineering student. Again, it was tedious calculations – this time the ones forming the bulk of his coursework – that prompted him to investigate the possibility of a machine to do the work for him. In the 1930s he worked on a number of different designs using electric relays and, notably, a binary numbering system. This seemed well suited to electric computation, as the binary system uses only ones and noughts, which can be readily expressed electrically as 'ons' and 'offs'. Zuse also worked out a way to input information into his computer using punched tape – in fact old movie film that he had modified. Through his work the four universal components that define a modern computer began to emerge in electro-mechanical form: a program or set of instructions, a central processing unit, a system of input and output, and a memory.

It was all going so well that in the 1940s he approached the German High Command for the funding to build a new computer, this time using the very latest valve technology. His request was turned down.

On the Allied side, the war was to lead to the creation of two very important computers. In England, inspired by the urgent need to break the German military communications codes, a valve-based computer, the

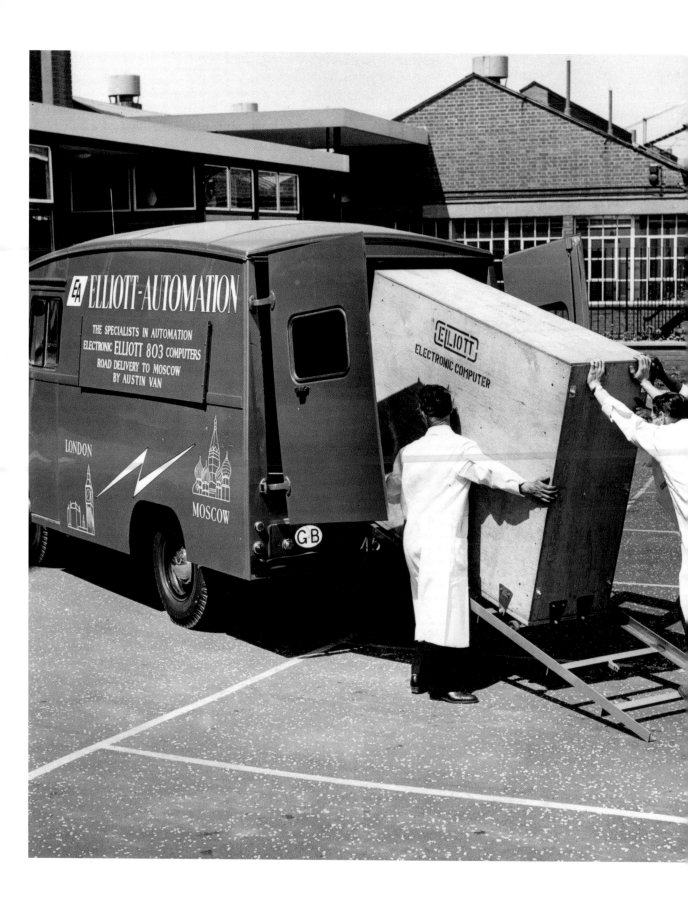

Colossus, was built. It was constructed at Bletchley Park, a country house ideally positioned to attract the nation's finest minds as it was situated roughly halfway between Oxford and Cambridge.

Colossus showed that, indeed, a computer using 2,400 valves could work. It was capable of processing 5,000 characters every second and was even reprogrammable, provided you didn't mind staying up all night with a soldering iron. The code the computer was charged with cracking was the Germans' Enigma, and in doing so the Colossus is believed to have shortened the Second World War by a year. The whole project was so secret that it was twenty-five years before the existence of Colossus was officially acknowledged. Bletchley Park's most renowned resident was Alan Turing, who is considered by most to be the father of computer science. He was later responsible for the first stored program computer (the ACE – the Automatic Computing Engine – which he designed while working at the National Physical Laboratory) and was a pioneer in defining artificial intelligence.

Not to be outdone, the Americans were building a much bigger and better computer on their side of the Atlantic. ENIAC, or the Electronic Numerical Integrator and Computer, was developed by the Ballistics Research Laboratory in Maryland. It weighed a massive thirty tonnes and relied on 18,000 valves, but it was 1,000 times faster than its predecessors. Its purpose was also military, to calculate the trajectory of shells, which it did admirably.

However, some reports suggest it usually worked for an average of only seven minutes, which was about the time it took for one of its many thousands of valves to fail. It was also prone to overheating. It was said to draw so much power that the lights in the local neighbourhood dimmed when it was turned on. But the most startling fact of all is that this computer, which was the size of a tennis court, was less powerful than a modern-day calculator. It is unlikely that anyone at the time predicted that within a few decades a computer might be fitted *inside* a missile.

If computers were to become more useful and affordable – and maybe even saleable to the general public – then, like radio, they would need to take the next step and leave the valve behind. Valves were a problem; the heat and light they generated even attracted moths, which led to constant breakdowns. So one job for the computer operative at the beginning of the day was 'debugging', and it is widely believed that this is where the notion of a destructive 'computer bug' came from.

So at the beginning of the 1950s computers said goodbye to the valve and turned to the new technology of semiconductors. These computers would be known as the second generation: still large and expensive, but now with reel-to-reel tapes spinning backwards and forwards, and with lights flickering seemingly at random. Computers were beginning to turn their increasingly powerful minds to non-military work; machines such as IBM's SABRE (Semi-Automatic Business Related Environment) computer which organised airline ticketing, or the UNIVAC machine that in 1952 predicted the result of the presidential election. However, because the outcome it gave – a landslide to Eisenhower – was so unexpected, those

SHARPLY POINTED PROJECTILE MOVING THROUGH THE AIR

$$(a^2 - u^2) \frac{\partial u}{\partial x} - u\,v\left(\frac{\partial u}{\partial y} + \frac{\partial v}{\partial x}\right)$$
$$+ (a^2 - v^2)\frac{\partial v}{\partial y} + \frac{a^2\,v}{y} = 0,$$
$$\frac{\partial v}{\partial x} - \frac{\partial u}{\partial y} = 0$$

IN THESE EQUATIONS, AT ANY POINT (x, y) IN SPACE

u = COMPONENT OF VELOCITY OF AIR IN x DIRECTION

v = COMPONENT OF VELOCITY OF AIR IN y DIRECTION

a, THE VELOCITY OF SOUND, IS GIVEN BY

$$a^2 = \frac{\partial P}{\partial \rho}$$

P = PRESSURE

ρ = DENSITY

ENIAC, another masterpiece of miniaturisation.

Above right:
Before ENIAC, solving these equations required huge teams of mathematicians with slide rules and log tables. Now it could be done by a machine as small as a house.

operating it re-programmed UNIVAC to give a more likely result. It turned out the operators were wrong and the machine's predictions had been correct all along.

In 1956 IBM introduced another breakthrough, the magnetic disc memory. The disc could hold a staggering 5 mb of data, or just enough to store one short music track or a decent photograph on today's computers. This state-of-the-art drive cost $50,000 but could be housed in a cabinet just five feet high – positively miniature by the standards of the time. But soon even individual transistors were old news and silicon chips, each effectively housing perhaps millions of transistors, took over. Although they were tiny, one chip could offer more processing power than the combined might of all the computers in existence at the end of the Second World War.

And it wasn't just the underlying technology that was changing, it was what we did with computers. By the late 1970s they were being found beyond factories, offices and military command centres; they were being given room in homes. In the early 1980s many people bought a Sinclair ZX Spectrum or a BBC Micro, even though they had no idea what to do with them. They had to be plugged into a domestic television monitor, could barely be used to write a letter, and certainly couldn't store music. Programs, such as they were, could only be stored on cassette tapes and loaded via a conventional player. None of this dissuaded enthusiastic owners from booting up their computers and running simple programs that allowed them to identify prime numbers or, at best, store a dozen phone numbers. It was all rather futile since recalling a phone number would involve plugging the various bits of the computer together and running the program, which could take at least three minutes and possibly an hour. But that wasn't the point. The computer had evolved from a recalcitrant engine occupying a whole room to something that could be accommodated on a desktop. It was never going to go away.

Sharp entrepreneurs quickly realised there were now people all over the world wondering just what to do with this thing they'd bought. The solution was to sell them computer games. Computer games were already becoming popular, particularly classics like Space Invaders which so far had been played in the company of a pint and a bag of crisps. In fact, there had always been a demand for a machine that you could play all on your own, thus eliminating the need for social contact. Even back in the eighteenth century there was a chess computer called the Turk that moved its own pieces. It was said to impress everyone, including Napoleon, but inevitably turned out to be nothing more than a scam: there was a man concealed inside. Computers, however, could take you on at chess for real. And at Space Invaders for that matter, and without the need to go out, run the risks associated with socialising, or even wear clothes. But in reality computer games were just a phoney front-room revolution, while in the shadows there was a real revolution going on.

If you ever attend one of the courses at SRI, the Stanford Research Institute, just south of San Francisco, you'll be schooled in the basics of what it takes to be an entrepreneur in California's Silicon Valley. They will tell you that you need to know the difference between work that is important and work that is merely interesting. The difference? Unless your idea is capable of generating at least a billion dollars then it isn't important. In Silicon Valley money talks and generally in the sort of numbers that only computers can deal with.

An important piece of work that must have seemed merely interesting (if not completely irrelevant) at the time was that pursued by Doug Engelbart. In 1945, when working as a radar technician, he read an article about the future challenges for science. It predicted that one day the world would be so packed with information that, if we were not to be overwhelmed, we would need a way to sort it and access it. Engelbart took up this challenge and went to SRI in the late 1950s. He worked on different methods for controlling a computer, and settled on using a pointer to move a cursor on the screen. At the time, computers were still a rarity and generally displayed nothing more than strings of code, so his idea was somewhat ahead of operational reality. Even so, by 1964 he had invented the mouse, and today it is estimated that you can find one on over 800 million computers. It was ideas like this and the GUI (Graphical User Interface) that were to transform computers from austere business machines to user-friendly home companions. Now, finally, they were ready to shrink the world.

Computers might have proved their prowess at long division, cracking codes or managing ticketing systems, but the idea that they might one day allow people all over the world to talk to one another was not on the agenda until the late 1960s. The first inklings that this might be possible came when two computers communicated with each other in 1969. It was an event that was almost entirely unnoticed. On 29 October 1969, a computer at the University of California, Los Angeles, sent a message to

'ELECTRONIC MAIL HAD EMERGED, THOUGH LARGELY UNNOTICED, AND WAS USED BY NO LESS A TECHNOCRAT THAN THE QUEEN'

another computer at SRI. The message was simple, just 'Log In'. Under pressure the computer fluffed its lines and crashed before getting to the letter 'g'. But what was said was less important than the simple fact that one computer spoke, and another heard. The notion of computers as sociable beasts had taken root.

The network that had linked the computers was the Arpanet. It was originally a military communications system, built by the US Defense Department and designed for the worst. In the 1960s 'the worst' meant nuclear war, so the Arpanet featured massive in-built redundancy. If any part of it was destroyed, information could be re-routed through another part of the network.

In 1969 just four host computers used the Arpanet, all funded, rather worryingly, by an organisation that couldn't even be relied upon to spell properly, the US Department of Defense. But as the threat of total obliteration from a nuclear attack diminished a group of American universities were allowed to connect up, and so, by 1971, twenty-three

computers were on the network. Other networks, run by businesses such as CompuServe or American Online, opened up to send messages. Electronic mail had emerged, though largely unnoticed, and was used by no less a technocrat than the Queen as early as 1976.

Fairly soon there were a great many different networks all operating side by side, but what was really needed was a way to network the networks. The result was the Internet, and by 1994 there were a staggering 30,000 sub-networks all connected up to this massive and essentially free global communications system. It was a miracle of happy co-operation, each individual network swelling an enormous and emerging infrastructure.

But, fast though this growth was, there remained the problem of searching all the information that was stored in the networks. It may have been the most comprehensive source of knowledge ever created, but it was about as accessible as a library of shredded books. The solution came in 1989 when Tim Berners-Lee, a young British scientist working at CERN, the European Particle Physics Laboratory, developed a way of organising the information so that it could be accessed across all the networks. The result of his work was the creation of the world wide web, which allowed the transfer of not just messages but ultimately music, photographs and even video. By 1996 there were 60 million on the web. By 2000, there were 250 million. Anyone with access to a reasonably modern computer could now delve into the biggest library ever created, and without having to apply for membership.

It is, undoubtedly, one of the defining innovations of the twentieth century. Yes, as critics are quick to point out, it can easily become a repository of completely duff information but, as Charles Babbage himself once said, 'Errors using inadequate data are much less than those using no data at all.'

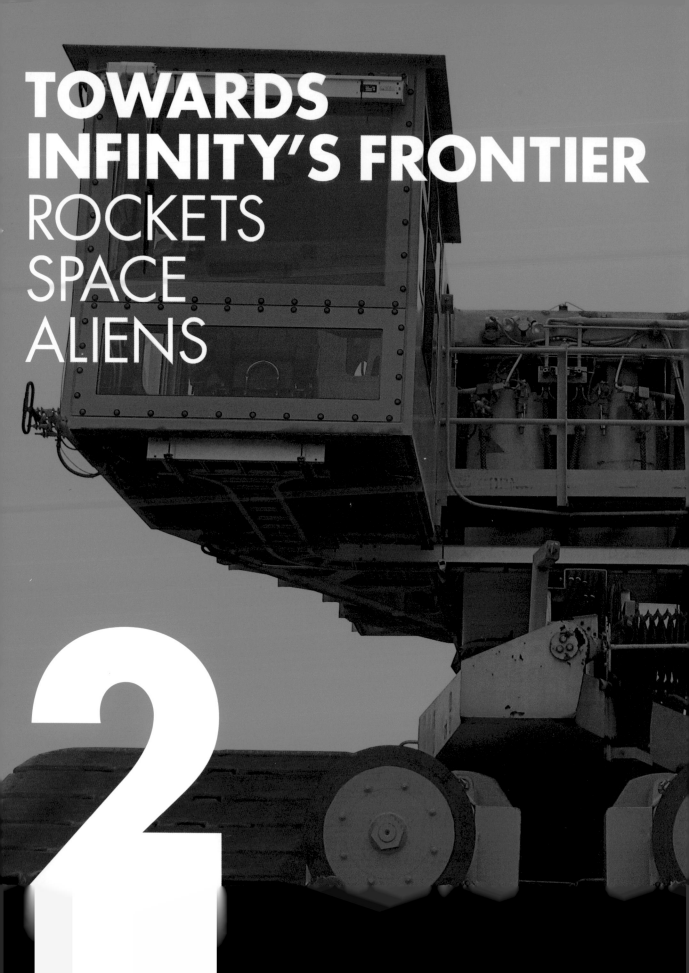

TOWARDS INFINITY'S FRONTIER
ROCKETS
SPACE
ALIENS

2

THE GREATEST RACE IN HISTORY

Astronaut John Glenn on board a mock-up of a Mercury space capsule. He would take the real one around the earth three times.

Of the twelve men who have walked on the moon, only nine are still alive and of those, the youngest is over seventy. The pioneers of the most unforgiving new frontier in the history of human exploration have no successors; they have founded no new colonies, and there are no towns named after them on other planets. Instead, they have assumed the status of veterans of an old campaign whose significance is fading and, like old soldiers, they too will soon fade away.

And when they are gone, an enormous irony will emerge. 'Going to the moon' will have slipped quietly from living memory and into history, and humankind will have retraced its one giant leap and arrived back at Earth. Is this not like circumnavigating the globe and then deciding that the world's roundness is an irrelevance?

Maybe not, because the most defining event of the twentieth century – a man walking on another planet – seems to have had remarkably little effect on our lives. Doubtless, it was a spectacular achievement and a stunning spectacle, and the wholesale embrace of space imagery can be seen to have inspired the design of everything from domestic door fittings to cars. The folklore, however, suggests that its most enduring legacy was the non-stick frying pan, which seems like a poor return for a programme that at one point was consuming around 4 per cent of the U.S. federal budget.

Elsewhere, at the Kennedy Space Center today, a machine vending rubber 'power balls' claims that they were the first product created in space. This is a great boast if you are twelve and you have one in your school blazer pocket, but even the best and most vigorously hurled power ball won't reach escape velocity.

The space race was a good thing, the standard riposte by the space lobby goes, because of the spin-off. But ask these space advocates to define spin-off in terms of actual day-to-day products, and they begin to mutter unconvincingly about the frying pan, sunglasses and Velcro. Few of them know about the power ball. A better argument is the one for the

President John F. Kennedy chooses to go to the Moon, 12 September 1962...

broader and less quantifiable effects that the space race had on fuelling the white heat of technology. The Apollo project, and those leading up to it, consistently broke new ground in engineering, manufacturing, systems design, logistics, human endurance and psychology. It depended on the ability to control and indeed overcome the forces of nature and the limitations of the human condition to an unprecedented extent. That, surely, left us with something?

And, as with all the best technical advances, the whole man-on-the-moon business was firmly rooted in the work of men in sheds. With time the sheds grew, and culminated in NASA's Vertical Assembly Building at Cape Canaveral, where the Saturn V rocket was assembled and where, these days, the space shuttle is mated with its external fuel tank and booster rockets. Until recently, when Boeing expanded its final assembly plant at Everett, this was the biggest building, by volume, in the world. But it is still a shed nevertheless, as the visitor will appreciate on entering and gazing up at 525 feet of sheet metal and exposed girder construction. It is also authentically chilly, and it is rumoured that without the air-conditioning system, small rain clouds would form in its rafters. Smells like a shed, too.

Along with the men in sheds, the other vital component of a space programme was cash. Anyone who has watched *The Right Stuff* will know what makes the ship go up – funding. To marshal the sort of sum needed required a man of great charm who also happened to be president: John Kennedy.

Kennedy was just forty-three years old when he was made thirty-fifth president of the United States. He was a man brimming with ambition for his country and he broke the mould of politicians for the time – young, energetic, charismatic and handsome. He was the first president in American history to be inaugurated without a hat on, an indication, perhaps, that his vision extended upwards as well as forwards.

He was also in a mess. Not long into his three-and-a-half-year tenure his foreign policy took a turn for the worse. One disaster led to another and finally to fiasco when he attempted to wrestle Cuba back from communism and its leader President Castro by invading the island, which left over 1,000 soldiers stranded on the beach at the Bay of Pigs. This cock-up was so bad that for years it was to be held up by sociologists as a fantastic example of how not to organise things. The nature of this failure was even given a name – 'groupthink' – to describe how well-educated people can collectively do the most dumb-ass things because no one has the courage to tell the boss his ideas are a bit pony.

Kennedy also faced growing unrest in South East Asia, a Cold War that was becoming distinctly colder, and racial tension at home. It's easy to see why he needed to come up with something a bit different and upbeat.

To put the lid on it, the Soviet Union, then America's arch rival, was already having some success in space. In 1957 they launched Sputnik, the first satellite, into orbit. Sputnik, not much bigger than a beach ball and weighing just 183 pounds, circled the earth every ninety-eight minutes emitting a regular beep that was, in truth, no more threatening than the chirp from a smoke alarm with a dying battery. But it scared the living daylights out of the Americans.

Then, on 12 April 1961, and whilst its great global adversary was still reeling, the Soviets launched a fantastically brave soldier into orbit. Yuri Gagarin went up for just under two hours, but that was long enough to complete one trip around the earth and, most significantly, overfly sizeable chunks of American airspace without permission and without fear of interception. Gagarin was not only the first man in space but also the first man to be promoted in space. He blasted off as a lieutenant and splashed down as a major.

For the United States, a country that had grown accustomed to being the first with everything, this was a huge blow to national pride; a poke in the ribs with a pointy space stick from its feared enemy. The people wanted a response, and the cost could go hang.

The National Aeronautics and Space Administration was formed in 1958 and on 5 May 1961 Navy Commander Alan Shepard Jr slipped out of the earth's atmosphere (though not into orbit) for fifteen minutes in his spacecraft Freedom 7. He was propelled on a Redstone rocket, essentially a piece of ordnance on which the warhead had been replaced by the tiny Mercury capsule.

...largely because he was worried that the Soviets might get there first.

'THE FIRST AMERICAN IN SPACE ARRIVED THERE NOT SO MUCH LIKE A SQUARE-JAWED HERO BUT LIKE A VAGRANT, LYING IN A POOL OF HIS OWN URINE'

Shepard was treated to an ecstatic New York ticker-tape parade by an adoring and reassured public. They may have been less effusive if they had known that the first American in space arrived there not so much like a square-jawed hero but more like a vagrant, lying in a pool of his own urine.

During the delay to the launch of his rocket, Shepard realised he would soon need to take a squirt. When a man's got to go, a man's got to go, whether that's into the new arena of space or merely to the john. Eventually, the call of nature become stronger even than the call of the new frontier, and NASA was faced with a unique decision: remove Shepard from the capsule and delay the launch even further, or allow him to do it where he sat, in the as yet untried spacesuit, surrounded by delicate instruments. Eventually, and to Shepard's relief, permission was granted to 'do it in the suit', and to NASA's relief the unpredicted increase in moisture levels did not cause any problems.

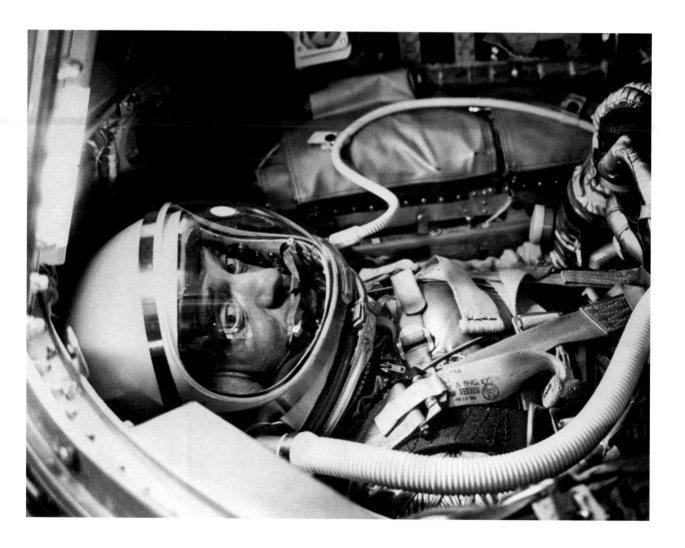

But they were determined not to be caught short again, and Shepard's
agony led directly to the 'space nappy' or, as NASA with its love of acronyms
preferred to call it, the MAG or 'maximum absorbency garment'.

Twenty days later, encouraged by the first American success in space
and quietly sidestepping the personal hygiene issue, Kennedy addressed
the American Congress and made his pledge: 'I believe that this nation
should commit to achieving the goal, before this decade is out, of landing
a man on the moon and returning him safely to the earth.'

It wasn't going to be cheap. He was asking for $9 billion at a time
when a smart family car cost around $2,500, but Congress accepted. The
true cost would turn out to be over $25 billion.

With only one slightly damp American in space and not even in orbit –
John Glenn would be the first to achieve that distinction a year later – the US
had made a commitment on a scale unequalled before or since. Why? To
throw down the gauntlet to the Soviets? To distract the American voter from
trouble at home? To indulge a world hungry for novelty? The following year
Kennedy explained exactly why: 'We choose to go to the moon. We choose
to go to the moon in this decade and do the other things, not because they
are easy, but because they are hard.'

And because, he might have added, as a spectacle, rocket launches
kick butt.

To watch a rocket launch is a unique and magical experience and,
as shock and awe go, is the real deal. To this day, the inhabitants of Cocoa
Beach, a few miles south of NASA's spaceport at Cape Canaveral, rarely
miss one. The beach was made famous in the 1960s as the playground of
the Mercury Seven, America's first astronauts. Now its white sand and
rows of low-rise bungalows offer the perfect view of the launch platforms
towering into the sky, four miles across the bay.

Since this is Florida, and the weather is invariably good, a party
atmosphere prevails. Barbecues are lit, beers are cracked open, space
geeks and lucky tourists mix with locals and keen observers from the space
trade. Radios are tuned to the chatter of NASA talk-back, obligingly laid on
by Mission Control, and expectant ears are cocked for the sound of the
start of the final countdown.

At zero, as the rocket ignites, launch virgins are astonished to see the
sky light up as though a second sun has risen at time-lapse speed. Upturned
faces are said to glow amber in the reflected rocket exhaust, even at a range
of several miles. All heads swivel upwards as one as man's ingenuity
confirms its dominion over the world's gravity in what can genuinely be
considered a blaze of glory. Around a minute later, the rocket is lost to sight.

The event is akin to the launch of a great ocean-going ship in the
nineteenth century. Even the most ardent Apollo 11 conspiracy theorist would
struggle to remain unmoved by the show. A launch, even though it is now a
relatively routine event – there have been over 100 shuttle launches to date –
remains the most stunning man-made phenomenon this planet has to offer. For
this peerless pleasure the Cocoa Beach crowd owes a debt to countless great
scientists but three in particular: one Russian, one American and one German.

Konstantin Tsiolkovsky was born in 1857, was very deaf and taught maths in his day job. He was also the first to make a serious stab at space travel, if only on paper. He claimed that humans would travel to other planets, and it wasn't just vodka talking on the long nights in the log cabin where he lived. He did the equations to prove it.

He calculated – correctly – that a rocket would have to travel at 18,000 mph to escape the pull of Earth's gravity and achieve orbit. He also proposed familiar solutions to many of the problems that space travel would come to present, such as multiple-stage rockets and even an airlock, should an astronaut fancy a stroll in space. All this, and the Wright brothers had yet to leave the ground.

In 1903 he drew his ideas together in a book, *The Exploration of Cosmic Space by Means of Reaction Devices*, where he suggested we should build space stations to help us explore our solar system. He even had an inkling of what fuel to use, favouring a liquid rather than a solid type. He was right about that, too, as an American, Robert Goddard, was to demonstrate.

Goddard had never heard of Tsiolkovsky when he began dreaming of flying to the moon. He took his inspiration from the giants of science fiction, Jules Verne and H.G. Wells. He dabbled with fireworks and gunpowder as a young man, but felt the answer lay elsewhere if he was to go very far into space. Solid fuels were hard to control, liquid fuels might not be. If he could mix hydrogen and liquid oxygen in a combustion chamber, then perhaps he would produce enough controllable power to go to the moon.

In 1920, in a report titled 'A method of reaching extreme altitude', he outlined his plans. But the newspapers, and in particular the *New York Times*, were having none of it. In an editorial they argued that there was no air to push against in space, so a rocket would be unable to propel itself forward. Deeply insulted, Goddard did what any true inventor would do, that is, not indulge in a legal wrangle or a battle of wits with the press, but spend two decades proving that his calculations were correct, and that a rocket would work in the vacuum of space.

In fairness, space had become a confusing topic. In 1887 Michelson and Morley had demonstrated that it was not, as previously thought, an absolute. Einstein's Special Theory of Relativity had been published in 1905. But the giant's shoulders on which the space race stood, even in the era of Apollo, were Newton's, and Newton, in his third law of motion, stated that 'For every action there is an equal and opposite reaction.' So, vacuum or not, if the rocket's exhaust exited one way, the rocket would accelerate the other way. Goddard understood this.

He built a few test rockets. One, called Nell, was about ten feet tall and was fuelled by oxygen and gasoline. It achieved a speed of 60 mph but rose only fourteen metres before nose-diving into a cabbage patch. Undaunted, he built other rockets, but it wasn't until he found a rich sponsor – Harry Guggenheim – that he could really get his teeth into this rocketry lark. Now he had sufficient funds to move to Roswell, New Mexico – later to become the hub of the famous aliens myth – and work undisturbed on his vision of spaceflight.

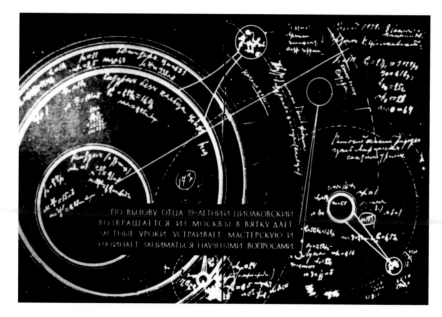

Konstantin Tsiolkovsky's calculation of escape velocity turned out to be remarkably accurate.

Right:
Robert Goddard with his – and the world's – first liquid-fuelled rocket, 1926.

Still Goddard's work was largely unappreciated, except by one small group of engineers who recognised his genius and read his papers voraciously. Unfortunately for space travel these men were German, so Goddard's work took a detour en route to the moon via Chiswick in the suburbs of London.

Stavely Road, Chiswick, looks more like a quiet suburban street of early 1900s housing than a turning point in the history of space exploration. But it was here, at 6.44 pm on 8 September 1944 that the first German V2 rocket struck. Because its flight was supersonic, no one heard it coming, and at first the destruction of a whole row of houses and the deaths of three people were blamed on an exploding gas main.

But no. Goddard's thinking had been refashioned into a missile that was now at the disposal of Adolf Hitler as the Vergeltungswaffe 2, the Vengeance Weapon number 2 (the first had been the ram-jet powered V1 'doodlebug'). The man behind the V2 was Wernher von Braun, the son of an aristocrat and the third genius to help put a man on the moon. To his credit, he was also a frustrated space pioneer, even during the war, and is alleged to have said, after the V2's first successful launch, that it had worked very well but unfortunately had landed on the wrong planet.

For now, though, the target was London. The V2 could carry a 2,000-pound warhead, rise to almost sixty miles above the earth, and hit its target at almost 2,000 mph after a journey of just a few minutes from its launch site in Amsterdam. Fortunately, the V2 was hard to aim accurately and did not prove decisive in the outcome of the war, but it did confirm von Braun and his team as the most experienced rocket scientists in the world. That meant that, at the end of the war, they were extremely valuable to both the Americans and the Russians.

The result was an unholy – and some say immoral – rush by the two emerging superpowers to secure the services of the German experts, regardless of a war record that might otherwise have seen some of them rotting in prison. In a fair world, the Germans would have been lined up and the Americans and Russians would have picked team members alternately, as in a school football game. The only difference would be that this time, those looking most like Brains from *Thunderbirds* would have been picked first, rather than the other way around.

But there was too much at stake for that sort of thing, so it was a smash-and-grab contest with no rules. Von Braun made a play to side with the US and take over a hundred of his rocket scientists with him. It seems that, whatever your past, if you could sketch a rocket motor from memory, you were in the team.

So America had bagged the boffins, but not without falling victim to yet more of the irony that characterised the race for space. Apparently, when US interrogators asked the German scientists how the V2 worked, they were met with incredulity. 'Just ask Goddard,' they said.

Still America was now equipped to begin the Mercury project, the first stage of the journey that would put a man on the moon within ten years of Alan Shepard's first sub-orbital foray. And there was never any doubt that it would be a man. Space was still a bloke's business.

There was a women's programme, run in near secrecy, that shadowed the selection of the Mercury astronauts. They were drawn from among the greatest women pilots in the US. The women became known as the Mercury 13 or the FLATs – First Lady Astronaut Trainees. But, good as they were – sometimes outperforming men in the battery of tests any prospective space pilot had to endure before selection – the FLATs were never to have their day.

A newspaper report 'reveals' the workings of the V2 terror weapon in 1944.

Jerrie Cobb, one of America's FLATs, or First Lady Astronaut Trainees, September 1960.

Right:
The Mercury Seven: Schirra, Slayton, Glenn, Shepard, Carpenter, Cooper, Grissom.

NASA at the time just wasn't comfortable with women, one chief going so far as to say, 'I would rather put a monkey into space than a woman.' In fact, the only woman to be allowed on to the gantry where the rocket was launched was the artist who painted the name on the capsule, and even then only because astronaut John Glenn insisted.

So men it was, and the Mercury programme intended to put seven of them in space in just two years. One, Deke Slayton, was removed from the roster after being diagnosed with a heart complaint – he later went on to become the Co-ordinator of Astronaut Activities for NASA and eventually, in 1975, piloted the American half of the historic Apollo-Soyuz joint US/Soviet space mission. Shepard's flight was followed by a second sub-orbital mission by Virgil 'Gus' Grissom. John Glenn and Scott Carpenter both completed three orbits each, Walter Schirra stayed up for six and finally, in May 1963, Gordon 'Hot Dog' Cooper flew twenty-two orbits to bring the programme to a close and cement America's lead in space. If the Christian names ring a bell, it's because they were the ones chosen by Gerry Anderson for the five Tracy boys who piloted Thunderbirds. Only Schirra failed to be immortalised in puppetry, perhaps because Anderson realised that a superhero named Walter was never going to cut it, even in a country where men are sometimes called Carrol.

The astronauts – and their wives, since all were married – became the most famous men in America. They signed deals with *Time Life* magazine and experienced the kind of celebrity only Elvis or the Beatles would understand; these undoubtedly brave men, fighter pilots all, spent their days training and enjoying the pleasures that their fame foisted upon them. They drank beer, bought powerful muscle cars and drove them flat out on the beach; cars like the Corvette and the AC Cobra. A trick employed by Cobra salesmen was to pin a large-denomination dollar bill on the dashboard and challenge the prospective customer in the passenger seat to reach out and take it under acceleration. No one could, the car created so much *g*. But what was that to a man who had agreed to being strapped to the top of a military-surplus rocket and fired into a world as yet unexplored?

To the surprise of cynics and doubters, the Mercury programme was concluded without losing a man in space. But within a few months, space had lost its champion on earth with the assassination of Kennedy on 22 November 1963. His vice-president, Lyndon Baines Johnson, was sworn in within three hours, Kennedy's widow Jackie at his side. Among the millions of questions asked about the new man was whether or not he would stay in the race.

The man who had sponsored the law that gave birth to NASA quickly made it clear that the goal had not changed. 'I do not believe,' he announced, 'that this generation of Americans is willing to resign itself to going to bed each night by the light of a communist moon.' In any case, by now the space race had fired the imagination of the US public and space fever was spreading across the globe. The money kept coming, the sun continued to shine on the astronauts, and the Mercury programme evolved into the Gemini programme.

'THE SATURN V WAS BIG, AND IT WAS CLEVER'

At 430,000 pounds, the Gemini rocket produced five and a half times the thrust of the Redstone that had allowed Shepard to poke his head above the atmosphere for fifteen minutes. Gemini 7 completed 206 orbits and successfully rendezvoused with Gemini 6, while Geminis 8, 9a and 10 successfully docked with unmanned Agena modules and Ed White performed the first EVA (extra-vehicular activity; a space walk to earthlings) from a Gemini spacecraft in June 1965.

All of this was excellent rehearsal for the operational activities that would be required of a moon trip, but the fact remained that all of America's manned extra-terrestrial activity to date had been conducted in orbit, at distances above the earth measured in mere hundreds of miles. But the moon at perigee (that is, the point in its path when it is closest to the earth) is about 225,000 miles away. Getting there would require a new set of skills that not everyone believed could be mastered. In the Mercury era, Hot-Dog Cooper confessed that no one had any idea how to navigate to the moon, and Gus Grissom went so far as to call the president 'nuts' for even suggesting it. And whatever else it took, the mission was going to need a very big rocket.

The Saturn V was big, and it was clever. Standing 363 feet tall, or roughly thirty-six storeys, it was assembled from over three million components and, through its three stages and eleven engines, produced 8,700,000 pounds of thrust. To this day NASA, in an uncharacteristic deviation from its usual acronymic habits, refers to it distinctly as a 'moon rocket'. Around 400,000 people were employed in its design, manufacture and operation, the din it created at launch is the second-loudest man-made noise ever recorded after the atomic bomb, and it remains the most complex and powerful machine ever built.

This is all the more remarkable when you consider *how* it was built. Space bores are fond of pointing out that the Apollo capsule boasted less computing power than the engine management system of a modern car. The design team had little more to play with. The Saturn V was conceived largely on drawing boards by the sort of men who keep a selection of pencils of varying hardness in clips on their breast pockets. It was built by America's foremost aircraft manufacturers using the tools, materials and techniques that were in place at the beginning of the jet age. Standing alongside the preserved (and horizontally displayed) Saturn V at the Kennedy Space Center, the visitor can marvel at a contraption of terrifying proportions that seems to have been contrived from domestic plumbing and air-conditioning supplies and cloaked in the tin-and-rivets skin of a Cessna light aircraft.

Once the Saturn V was complete inside the massive Vertical Assembly Building, the public had plenty of time to admire it. Still vertical, it made the three-and-a-half-mile trip to the launch pad at less than one mile per hour atop the 'crawler', a multiple-tracked vehicle that would occupy more than half a football pitch. Since the rocket had to be held to within one degree of vertical to remain stable, the crawler incorporated a self-levelling system that required constant adjustment on the journey, especially on the five-degree incline leading to the pad. The same machine, with a few modifications, is

The start of the first manned moon mission: 16 July 1969 and the launch of Apollo 11.

still in use today as a transporter of assembled space shuttles.

The Saturn V and every system and structure attendant on it showed NASA at its most potent. It was breathtaking audacity, a truth more stunning than any space fiction stemming from Gerry Anderson's imagination. The Saturn V team had redefined the boundaries of engineering and rewritten large chunks of the *Guinness Book of Records* even before the engines had been fired.

To sit at the top of a Saturn V waiting to be launched required superhuman courage. The G-forces experienced soon after take-off were enormous and made the AC Cobra feel like an uphill bicycle ride. The chosen few who flew the Saturn knew the risk they were taking; they were largely military pilots and had been to more funerals than they cared to remember. In the event of a problem on the launch pad, there were means of escape down the supporting tower. Within the first few seconds of flight the Saturn's communal ejector seat – essentially a small rocket attached to the command module to pull it free and allow it to float to earth – could be used, but after that there was no escaping the fury of the giant beast. It can not have been of much comfort to the three astronauts to think that, once they were strapped in and ready for blast-off, the next nearest people were required to stand at least three miles away.

The astronauts were dismissed by some as just 'spam in a can', and it was suggested that they were merely doing a job that a monkey had already done. The difference, though, was that the monkey didn't know it was sitting on top of a gigantic potential bomb. When Fritz Lang created the launch countdown to add drama to his 1931 film *Woman to the Moon* he probably didn't realise that one day the wives and families of real astronauts would find it terrifying.

But it wasn't a launch that claimed the lives of the first three Americans to die in a spacecraft. It was a static test. In January 1967, during pre-flight checks for the planned first manned Apollo mission, using the smaller Saturn 1B rocket, a freak electrical spark triggered a flash fire in the oxygen-rich environment of the command module, killing astronauts Virgil Grissom, Edward White and Roger Chaffee in seconds. It shook NASA's confidence badly, and reminded the public how hazardous this space business was and how heroic its pilots. But after an investigation of just two and a half months, the Apollo programme continued.

By December 1968 NASA was back in its stride and the Saturn V, with only two unmanned launches under its belt, was deemed ready to take the men of Apollo 8 to the moon.[1]

Not to land on it, but to fly within a few miles of its surface, where the weak lunar gravity would capture their spacecraft and hold it for a planned ten orbits. There was a big proviso attached to the plan, though; namely, that the calculations were spot on. One small miscalculation on a slide rule and Frank Borman, William Anders and James Lovell (later the commander of the ill-fated Apollo 13) would shoot past the moon and be lost to infinite space for ever, watching the earth dwindle to a speck until their oxygen ran out. It's not as if anyone was in a position to come and rescue them.

But on Christmas Eve 1968 the arithmetic was proved gloriously accurate and the astronauts found themselves a tantalising seventy miles above the lunar surface. As the three men swept around moon they transmitted the most remote and thought-provoking message in the history of civilisation: a reading from the book of Genesis, the story of earth's creation recited by men engaged in the greatest event since the one they were recounting and as far from the home of humankind as anyone has yet been. While they were there, they also took some holiday snapshots, including the image now known as 'Earth Rise'. It's still considered the definitive portrait of our planet.

So now the rocket, the spacecraft, the systems and the navigational techniques were in place and ready for Kennedy's vision to be realised, ahead of schedule if not quite under budget. On 20 July 1969 Neil Armstrong and Edwin 'Buzz' Aldrin steered the lunar module of Apollo 11 (named 'Eagle' after the bird on the craft's insignia) to the surface of the moon, while Michael Collins waited overhead in the command module 'Columbia', named after the spaceship-launching cannon in Jules Vernes's novel *From the Earth to the Moon*.

July 1969, Apollo II's lunar module leaves the moon to begin the long journey home.

Overleaf:
The first view in history of the earth in its entirety, taken from Apollo 8 in December 1968.

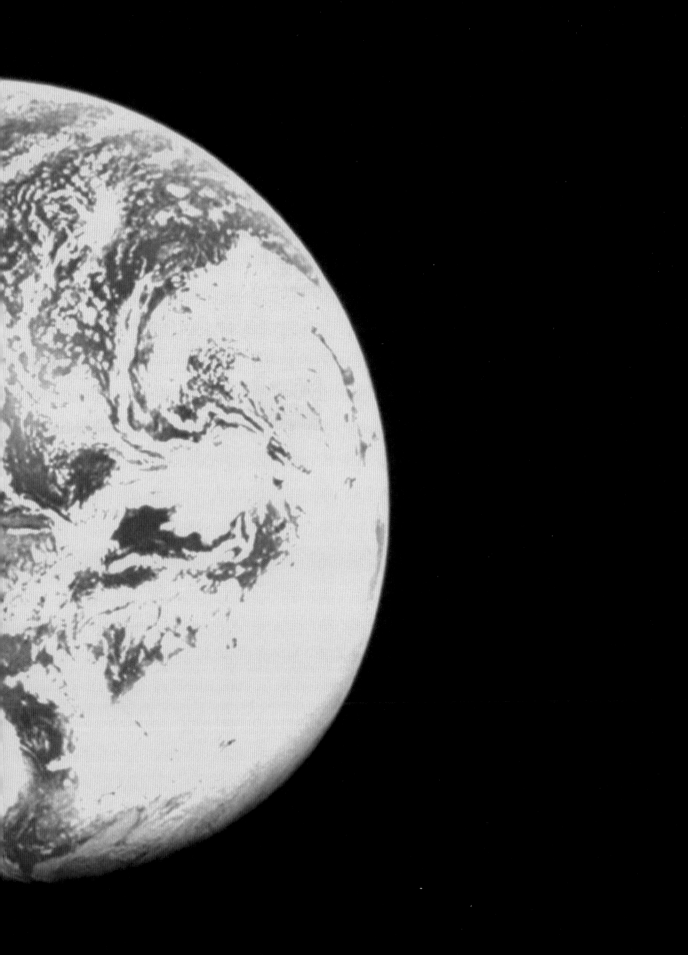

Armstrong and Aldrin were long on nerve but terrifyingly short on fuel. During the descent to the surface they realised that the module's automatic systems were going to deposit them in a rock-strewn crater several miles from the intended landing spot. Armstrong took manual control and, helped by guidance and encouragement from Aldrin at the radar screen, landed with less than 30 seconds' worth of fuel to spare.

It was a demonstration of skill and cool never equalled. If confirmation were needed that these men had the right stuff, Armstrong had accepted the mission despite rating their chances of a successful moon landing at 50 per cent. Unbeknown to him, Collins, alone in the orbiting command module, had privately given them the same odds. A TV audience estimated at 600 million saw that the Eagle had landed safely.

NASA had scheduled a period of rest for Armstrong before he stepped onto the lunar surface. He was meant to sleep – sleep, with the face of the moon at the end of a short ladder! – but instead he prepared a few words to mark the occasion, despite being genuinely unaccustomed to public speaking and having bated the breath of the biggest live audience a single man has ever had. Today that job would be given to hordes of PR executives or political speech-writers to ensure that all the sponsors were mentioned, but few people could improve on Armstrong's modest 'one small step for a man, one giant leap for mankind' moment. Some say he fluffed the most important announcement in human history and missed out the 'a', but more recent audio analysis suggests that he got it right. In any case, it seems unlikely that anyone was picking holes in the dialogue. A man was standing on another planet, and he would return home safely as the president had demanded. This, surely, was the crowning achievement of the century.

So what was its legacy? Certainly not the non-stick frying pan of popular mythology, because that was already in the shops well before Kennedy made his appeal to Congress. The cordless power drill, powdered drinks and smoke alarms can all lay some claim to having been spawned by the space race, but the cynics have much better material to work with.

They can point out that between Apollo 11 and the sixth and final moon landing in 1972, public interest waned and the television audience dwindled to the point where some of the programme was deemed not worth showing; that the Sea of Tranquillity proved rather too quiet for comfort and that the moon mission was not the instigator of widespread interplanetary travel but merely an elaborate and expensive way of confirming that it wasn't worth the effort; that the legacy of the race to the moon is a few unsightly piles of space junk gathering moon dust and three badly parked cars.

Maybe Kennedy had the right idea, that it should be done because it is hard, and that the ambition, the application and the pioneering urge would bequeath an example to the world. As Michael Collins of Apollo 11 later said: 'It's human nature to stretch, to go, to see, to understand. Exploration is not a choice, it's an imperative.'

But it seems that the final irony of the race to the moon – the first milestone on the road to the stars, as Arthur C. Clarke put it – is that it engendered a deeper love and appreciation of home. Earth, when viewed as a whole and as just one coloured dot in the vast canopy of space, seems to scream its fragility to those who have been privileged to see it that way. As the writer and world peace advocate Norman Cousins put it: 'What was most significant about the lunar voyage was not that man set foot on the moon but that they set eye on the earth.'

But the sentiment is better expressed by a wide-eyed and astonished Neil Armstrong, the first man on the moon: 'It suddenly struck me that that tiny pea, pretty and blue, was the earth. I put up my thumb and shut one eye, and my thumb blotted out the planet earth. I didn't feel like a giant. I felt very, very small.'

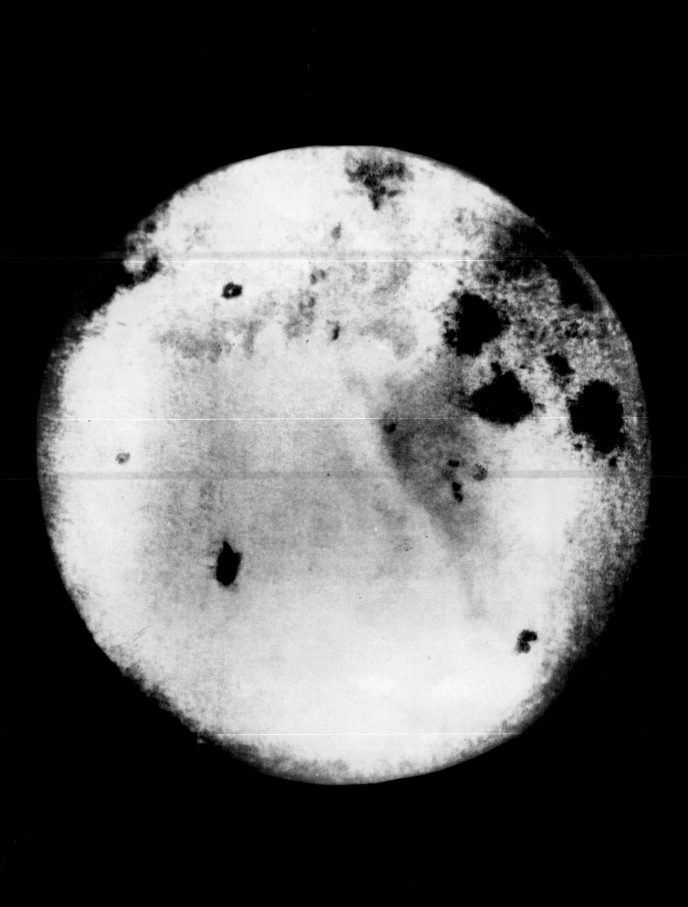

OVER THE MOON

The far side of the moon, as photographed
by the Soviet Luna 3 in 1959.

There was a very apparent side effect of the Apollo 11 mission: a severe
outbreak of moon fever. The moon had been shown to be within reach and
so, reasonably enough, people were starting to plan their own trips into
space. The idea had been put in everyone's mind the previous year when
Pan Am's president, Juan Trippe, called ABC-TV to make a bold
announcement: his airline would begin taking reservations for future flights
to the moon. The following day, the *New York Times* reported that the airline
had been deluged by eager prospective moon tourists and so, taking further
advantage of this heaven-sent PR opportunity, Pan Am created the Moon
Flights Club. In reality, membership turned out to be nothing more than the
longest check-in delay in the history of holiday-making.

But to be fair, the idea of a space vacation wasn't as far-fetched as it
now seems. NASA had already announced its intention to build a colony on
the moon by the mid-1980s, and if anyone was going to fly there it would be
Pan Am. In the popular imagination, they had already started, since the
spacecraft seen ferrying passengers to the moon in Stanley Kubrick's *2001
– A Space Odyssey* was clearly a Pan Am machine, complete with its
conspicuously large blue ball logo. The moon had never seemed closer.

Over 93,000 people wrote to Pan Am to reserve a seat on the upcoming
service. It would be nice to think that a cynical wife somewhere said to her
misty-eyed, moon-struck husband, 'You're more likely to become the president
than go to the moon.' It would be especially nice if the cynical wife was one
Mrs Reagan, since in the case of her husband Ronald it turned out to be true.

The third and perhaps most depressing effect of the Apollo 11 mission
was that it would come to stand as the most crushing indictment of the fickle
nature of humankind. Public fascination with space waned very quickly.
It was apparent after just four months. When Apollo 12 set off to make the
second landing, there just wasn't the same sense of global excitement.
Everyone can name the first two men to walk on the moon, but who knows
that the second two were Pete Conrad and Alan Bean (the pilot of the

command module was Richard Gordon)? Or that they landed on the Ocean of Storms? Or that there was a moment of terror at Mission Control soon after the launch, when a bolt of high-altitude lightning momentarily knocked out the spaceship's telemetry? What Kennedy had described as the greatest event since creation was already looking a bit old hat.

Until, that is, Apollo 13.

The second most famous phrase ever uttered in space is also the most consistently misquoted, perhaps because it has been remodelled as a sort of paradigm for how to remain calm and collected in a crisis. What was actually said, first by the command module pilot John Swigert, was 'OK, Houston, we've had a problem here.' Commander James Lovell then repeated the transmission: 'Houston, we've had a problem.'

And it was a big one. As the crew operated the electric motor used to stir the oxygen in the service module's number two tank, an electrical fire caused by damaged wiring overheated its contents, raising its pressure until the point where the tank exploded (this was established later. During the flight, it was thought that a meteor strike might have caused the damage). Critically, the explosion also damaged the number one tank.

The oxygen that had been lost was used to generate electricity for the service module, which was now left very low on power. Back-up batteries were available in the command module, but these gave only around ten hours' worth of power, all of which would be needed for re-entry. The crew had no option but to move into the lunar module and us it as a 'lifeboat', a procedure that, fortuitously, had been devised not long before the mission.

Apollo 13 was approximately 200,000 miles from earth and heading away from it at a rate of 20,000 miles an hour. Its crew were in danger of becoming the first people to be lost in space, which would make the

Houston – we've found some cardboard. On Apollo 13 mission control improvised a life-saving solution for the astronauts in space.

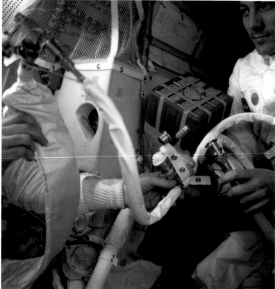

disappearance of Burke and Wills in the Australian Outback look like a
mishap in the garden.

The lunar landing was clearly off, but a trip around the moon was still
very much on. To expedite the craft's return to earth, it was decided to
increase speed and use the slingshot effect of the moon's gravity during one
pass to accelerate the inbound leg and shave ten hours off the journey time.
But even this plan was not without risk. Normally, it would involve firing the
service module's propulsion engine, a routine operation. But no one knew
if had been damaged in the explosion and so, after extensive debate at
ground level, the manoeuvre was effected by using the lunar module's
descent engine. It was Alcock and Brown all over again, a seat-of-the-pants
solution to a problem that no one had ever encountered before. Perhaps
this was what Kennedy had meant when he said, 'Those who dare to fail
miserably can achieve greatly.'

The astronauts were still by no means out of danger. The lunar module
was designed to sustain two people for two days. It was about to become
home to three people for four, and the lithium hydroxide canisters used to
scrub the internal air of carbon dioxide would not last that long. The
command module had its own scrubbing system and a plentiful supply of
spare canisters but, as if the whole thing had been devised as an object
lesson in the benefits of standardisation, they wouldn't fit the system in the
lunar module. Under instruction from Mission Control, and following
extensive experimentation on the ground, the crew used scrap materials
such as cardboard to fashion a new receptacle for the 'wrong' canisters.[2]

That wasn't the end of it. More improvisation was needed to preserve the
craft's dwindling power supply to ensure that enough would be left for re-entry.

Moon rock collected on the Apollo 15 mission, seen here in the hands of Jack Schmitt, later of Apollo 17, seen below leaving the moon.

In conserving it, temperatures on board dropped to dangerous levels, causing condensation. This, it was feared, would cause a malfunction when electrical systems were re-energised for the return home. So even with the crew on course and breathing safely, there was still no guarantee that they would come out of it alive.

Now the world was once more riveted by spaceflight. The power problem meant that no broadcasts could be made from Apollo 13, so news of its agonised progress through space was relayed to the world using diagrams, models and explanations from engineers at NASA. At the time of its successful splashdown, it is estimated that a billion people were watching around the world; more than had tuned in for Armstrong's historic 'small step'.

Clearly, simply 'going to the moon' was not enough to keep the public entertained, and NASA needed a new turn. They had one ready: the Lunar Rover Vehicle (yes, the LRV) or 'moon buggy' as it would almost instantly become known. After all, self-respecting Americans didn't like walking that much, even given the rare privilege of a whole planet to themselves.

So Apollo 15 blasted off, like some interstellar Winnebago, with a car on board, allowing James Irwin and David Scott to take a recreational spin around the dead planet once they had set up camp. The LRV is certainly the most expensive and exclusive car ever built. The original contract was for four vehicles at a cost of $19 million; inevitably, they ended up costing $38 million, which didn't even include delivery. It was a true all-wheel-drive machine, with a separate electric motor for each wheel producing a total of 1 hp. Top speed was 8 mph if you put your moon boot to the boards (three were used on moon missions, the fourth was cannibalised for spares).

With its balloon tyres and beach-buggy-from-outer-space styling, the LRV was a huge hit with the public, especially when it was seen bouncing around the lunar surface kicking up a slow-motion rooster tail of moon dust. That the dust covered the astronauts, obscured the buggy's control console and interfered with communications equipment was neither here nor there. Americans were not only walking on another planet, they were driving on it, too. John Young, the commander of Apollo 16, even managed to crash his LRV, though only slightly.

It wasn't all fun and games, though. The LRV was designed to allow the astronauts to travel further in their quest for revealing rock samples. Without it, said Harrison Schmitt, 'our current understanding of lunar evolution would not have been possible'. In total, the Apollo moon missions brought home 381.7 kg of rock and 'other materials'. Ominously, the rocks had been affected by many impacts; a stark reminder of what a good job the earth's atmosphere does in burning up dangerous space projectiles before they do any damage.

Originally, there had been moon missions planned as far ahead as Apollo 20, but a drastically shrinking NASA budget, plus the decision not to produce a second batch of Saturn Vs, put a stop to that. In 1972, Apollo 17 made the last manned visit to another world, and it was left to Eugene Cernan to turn out the lights on the greatest voyage of discovery in history. The last thing he did was to park the LRV so that its on-board

'CLEARLY, SIMPLY "GOING TO THE MOON" WAS NOT ENOUGH TO KEEP THE PUBLIC ENTERTAINED, AND NASA NEEDED A NEW TURN'

camera would broadcast a good shot of the lunar module's departure, and scratch his daughter's initials in the dust so that they would remain undisturbed for, as he put it, 'more years than anyone could imagine'.

The astronauts who had been involved were left to get on with their earth lives, which for some meant turning to alcohol, religion, parapsychology, art and celebrity. Life after a moonshot was clearly troublesome. On earth, the furthest anyone could physically be from home is about 12,000 miles. Viewing the planet from over a quarter of a million miles away seemed to trigger despair and frustration in some who had experienced it.

But what of the Soviet Union? Its impressive list of space firsts is often overlooked: first dog in space (Laika in 1957), first man in space and in earth orbit (Yuri Gagarin in 1961) and first woman in space (Valentina Vladimirovna Tereshkova in 1963). The Soviets were the first to achieve escape velocity when, in 1959, the spacecraft Luna 1 was successfully flown past the moon before entering an orbit between earth and Mars (the Luna 1 mission also marked the first time a rocket was successfully fired in orbit). In the same year, Luna 2 impacted the moon's surface and Luna 3 recorded and transmitted the first images of the dark side of the moon (the Soviets also launched the first probes to Mars – Marsnik 1, 1960 – and Venus – Venera 1, 1961).

But the Russians had long since accepted that they would be beaten in the race to walk on the moon, even though they had built the Lunniy Orbitalny Korabl (or LOK, their version of NASA's command module) and the Lunniy Korabl (or LK, a lunar module) and tested them successfully in earth orbit. Only the failure of the giant N1 rocket had held them back. They had even chosen their man to be first on the moon; or rather, he had effectively selected himself. His name was Alexei Leonov.

Leonov was one of the original twenty air force pilots selected to be the first cosmonaut group in 1960, the Soviet equivalent of the Mercury Seven. In terms of style and demeanour, he was a radical departure from the clean-cut, clean-living and slightly nerdy image the American astronauts presented to the world on the covers of a thousand magazines. Leonov was a bear of a man, with a physique that seemed to have come not from the gym but from toiling in the fields like some archetype of the glorious Russian peasantry. He was also an accomplished painter who made sketches in space and then used them as studies for larger works in his studio.

And although the honour of being the first man to walk on the moon eluded him, another was already his. On 18 March 1965 he left the relative safety of his spacecraft Voskhod 2 and pushed himself out into the icy infinity on the end of a safety tether to perform the first space walk.

All went well for the first ten minutes. But then Leonov's pressurised suit began to balloon like Bibendum's outfit in the vacuum of space. He became so big that he couldn't possibly re-enter the hatch to the spacecraft; neither could he effectively grip anything from within the swollen spacesuit. Fully five years before the drama of Apollo 13, he was in danger of becoming the first man to be lost in space.

Laika, the first dog in space and the first to die there.

Below:
Valentina Tereshkova, the first woman in space, 16 June 1963.

Opposite:
Alexei A. Leonov, hero of the Soviet Union and the first space walker. He carried a gun in orbit to defend himself against bears.

Overleaf:
The only way to service the Hubble Space Telescope – in space.

At the risk of unconsciousness, which would ensure that dubious distinction, he opened a valve on the suit to bleed off some of its air. In fact he had to reduce the pressure before he could squeeze his huge frame back through the air-locked hatch of Voskhod 2, head first. He was declared a Hero of the Soviet Union (an image of him walking in space appeared on Soviet postage stamps).

The whole mission was plagued with technical problems, including a troublesome re-entry that deposited the capsule hundreds of miles off target in the Ural mountains, where Leonov and his crewmate Pavel Belyayev had to spend a night surrounded by wolves. In an interview for the BBC Leonov revealed that he carried a loaded gun on board the spacecraft. It was assumed that this was in case he needed to end his life quickly in the event of a disaster in space. He laughed at the suggestion. It was in case he needed to shoot bears. But he did admit, in a much later interview, that he carried a suicide pill in his capsule.

Significantly, Leonov later said, on the subject of space walking: 'Building manned orbital stations and exploring the universe are inseparably linked with man's activity in open space. There is no end of work in this field.' The Soviet Union may have been beaten to the moon but it was ploughing another, different and some would say ultimately more beneficial furrow – a permanently manned space station, and therefore a constant presence in space.

And so, in April 1971, the Soviet Union clocked up another space first with the launch of the orbiting space station Salyut 1. The first attempt to board it, by the crew of Soyuz 10, failed because of a docking problem. The crew of Soyuz 11 were more successful, remaining on board for twenty-three days, only to die on re-entry to the Earth's atmosphere owing to a fault with a pressure equalising valve. Despite this, Salyut 1 was the first of seven progressively more complex such stations (eight, if the enigmatic Cosmos 557 is included). Cosmos 557's launch was a failure, so its true purpose was kept hidden from the observing Americans and it was allowed to burn up in the atmosphere. It was later revealed to have been a Salyut station. These stations were operated with varying degrees of success and some fulfilled a distinctly militaristic role. More famously, but largely because of the way it would eventually return to earth, America launched Skylab in 1973.

Leonov's second trip into space was to be, as he would have wished, inextricably linked to the development of the space station, even if it wasn't obvious at first. Leonov was the commander of Soyuz 19, which, on 17 July 1975, successfully docked in orbit with the last of the Apollo spacecraft, commanded by Tom Stafford.[3] Three hours after docking the two commanders exchanged the first international handshake in space through the hatch of the Soyuz. The original plan had positioned this event above the British seaside town of Bognor Regis, but a short delay, as if instigated to uphold King George V's dying instruction to 'bugger Bognor', meant it took place above the French town of Metz.

The union of the two spacecraft lasted for forty-four hours, during which time the Americans and Russians exchanged flags and gifts of tree

seeds, which were later planted in the two countries. They also signed certificates, visited each other's spacecraft, and attempted each other's language. Tom Stafford had a pronounced cowboy drawl when speaking Russian, which led Leonov to remark that there were three languages spoken on the Apollo/Soyuz mission: Russian, English and Oklahomski.

The mission was considered a symbolic one. The Apollo era was at an end, so Stafford's spacecraft, launched atop a surplus Saturn 1B rocket, was not given the expected sequential Apollo number (some historians have designated this launch Apollo 18, but NASA is having none of it). It was intended to celebrate the whole programme. The moon race was effectively at an end, too – at least as far as manned journeys were concerned – so Apollo/Soyuz signified a new spirit of collaboration, the first sign of a thaw in the Cold War perhaps. More importantly, the docking procedure was a test of the two nations' ability to make their spacecraft compatible, and its success would have far-reaching consequences.

America would not go to space again for six years, and when it did, it would be in a different type of spacecraft. On 12 April 1981, some twenty years after Gagarin rubbed America up the wrong way by orbiting the earth, NASA launched Columbia, its first STS, or Space Transportation System. Because it was reusable, like no other spacecraft before it, the more usual name of 'space shuttle' suggested itself. Had it been available on schedule – that is, in 1979 – it would have been used to 'repark' Skylab in a higher orbit. Since it wasn't, Skylab fell to Earth.

The Soyuz SL-4, the rocket that built the MIR space station.

Deke Slayton and Alexei Leonov meet in orbit, July 1975.

Overleaf:
The world's only successful reusable spacecraft, NASA's space shuttle, ready for lift-off in 2001.

Two things distinguished the shuttle from any previous spacecraft. Although it blasted off, as Apollo had done, vertically (and indeed from the same launch site), and shed its booster rockets and main fuel tank in flight, the ship itself could return to earth to land like a giant glider and be used again. It could also carry a decent payload – 22,700 kg of space luggage – and a crew of up to eleven people (eight is the current record).

The shuttle story would be marred by two tragedies. On 28 January 1986, Challenger exploded just one minute into its flight, killing the five men and two women on board. Until that moment, in twenty-five years of space travel, seven people had been lost. In an instant, the number doubled. One of the crew was Christa McAuliffe, a school teacher destined to be the first civilian in space. She'd been selected from over 10,000 hopeful applicants.

The cause of the disaster was eventually traced to a faulty 'O' ring in one of the solid rocket boosters, the elasticity of which had been severely compromised by the sub-zero temperatures at the time of the launch. Many engineers had warned against the dangers of launching in such conditions. The investigation, and subsequent modifications, kept the shuttle grounded for almost three years.

On 1 February 2003 a second shuttle, Columbia, was lost, this time on re-entry after a sixteen-day mission. Another seven-man crew perished, and the shuttle programme was again delayed, this time for two years.

But even before the first tragedy, the shuttle had proved its worth. As a means of launching military and civilian satellites, it was without equal. It was even employed, like a breakdown organisation's van, to repair the Hubble telescope in orbit and save it from an embarrassing attack of myopia. It could not go to the moon, but operated superbly in orbit. And that, as it turned out, was where all the manned space activity would now be concentrated.

Meanwhile, the Soviets were pursuing their interests in space stations. Following the success of the Salyut series, on 19 February 1986, they launched the first module of the MIR space station. Given the Soviet aim of a permanent presence in space, it is no surprise that this first module was the station's living quarters.

Six more modules were added to MIR between 1987 and 1996. All assembly took place in orbit, confirming what Leonov had said about the importance of 'man's activity in open space' – during MIR's service life, seventy-eight space walks were made for the purpose of assembling and inspecting it. MIR holds the record as the host craft for the longest continuous human presence in space, at just eight days short of ten years, and one of its many crew members, Valeriy Vladimirovich Polyakov, holds the record for the longest stay in space at 437 days. Before it was deliberately allowed to break up in the earth's atmosphere on 23 March 2001, MIR had completed 89,067 orbits.

Yet while the Soviets were definitely ahead of the game in space stations, they were behind it in space flight. Most of the MIR modules had been launched on Proton rockets and were serviced by Soyuz spacecraft and Progress cargo ships. Their own Buran shuttle had been abandoned owing to escalating costs.

Cosmic construction worker David Wolf performing an EVA (extra-vehicular activity) during assembly of the International Space Station in October 2002.

The same would come to be true of America's answer to MIR, Space Station Freedom. Announced by Reagan in 1984 and constantly revised over the next nine years, it was never built, eventually being dismissed as financially unviable in 1993. So the Soviets had the successful space station and the Americans had the means to service one effectively. Neither nation could afford both, but the fall of communism in 1991 inspired a continuation of the co-operative spirit engendered with the Apollo/Soyuz mission. More importantly, Apollo/Soyuz had shown that spacecraft from the two superpowers could coexist happily.

On 29 June 1995, under the command of Hoot Gibson, and travelling at less than one inch per minute, the shuttle Atlantis docked with the MIR space station. It was the hundredth manned US space launch, and the union of shuttle and station created the largest man-made object in orbit,

James May's 20th Century: Over the Moon

Dennis Tito, the world's first space tourist.

with a combined mass of 250 tons. US astronauts would spend over two years on MIR, learning from the Russians' long experience of sustained space dwelling.

The shuttle/MIR collaboration was also something of a political feeler; a test to see if the two old rivals really could live happily together in space. There were tense moments, language barriers and moments of distrust, not to mention a couple of near disasters; first a fire, and then a collision between MIR and an unmanned spacecraft, both in 1997. By 1998 the US and Russia, along with the space agencies of Japan, Canada and Europe, were ready to embark on the construction of the new International Space Station, or ISS.

The first section of the new station to be launched, the Zarya Cargo Block, was delivered in November 1998 on a Russian Proton rocket. Two further pieces were added before the first crew moved in on 2 November 2000. The ISS is an ongoing project and, like MIR, is being assembled in space. It combines elements of the aborted American Space Station Freedom, Russia's planned MIR 2, Europe's Columbus Space Station and Japan's Experiment Module. It has been designed to be a place where astronauts and scientists can work in a spirit of international co-operation.

Or has it? As the twentieth century drew to a close, a third era of space had arrived, that of the space tourist. The idea of taking fee-paying passengers into space had been part of the Soviet plan for the original MIR station. Dennis Tito, an American businessman, reserved a trip, and when MIR was decommissioned he opted to re-route to the ISS instead. On 28 April 2001 he became the first space tourist with a ten-day round trip to the ISS. At the time of writing, three others have followed him.

Tito's ticket cost him in excess of $20 million, but the price of an away-break in space could come down considerably. Virgin Galactic are quoting $200,000 for a planned two-and-a-half-hour space trip. The Hilton chain has suggested putting a hotel in space and fashion designer Eri Matsui has designed clothing, including a wedding gown, intended to look its best in a weightless environment. It may be only a matter of time before the sanctity of low Earth orbit is completely ruined by silly hats and cheap lager.

On the other hand, it has been suggested that NASA's replacement for the shuttle, the Crew Exploration Vehicle, could be available for fee-paying trips to the moon as early as 2020. But you may have heard that somewhere before.

WE KNOW YOU'RE OUT THERE. DON'T WE?

Left:
This isn't a real alien, rather like all the others aren't.

Below:
On the other hand, this probably isn't as well. Two US airforce officials identify fragments of a crashed weather balloon at Roswell, New Mexico, in 1947. Obviously this was a cover-up; in truth it was a spacecraft full of ALIENS!!!!!

'I am convinced that these objects do exist and that they are not manufactured by any nation on Earth. I can therefore see no alternative to accepting the theory that they come from some extraterrestrial source.'

Anyone who considers that the search for flying saucers and alien life is the preserve of nutcases and cranks might like to consider that the above words were penned by no less a man than Air Chief Marshal Sir Hugh Dowding. It was he who masterminded the RAF's victory in the Battle of Britain. He was hardly a lunatic and, famously, was not given to histrionics.

In the twentieth century the alien came of age. He, she, or it had existed before, of course, most famously in H.G. Wells's novel *The War of the Worlds*, which was published in 1898. But the twentieth century provided the material for stoking enthusiasm (or fear, depending on which way you viewed these things): on the radio where Wells's book was dramatised in 1938 by Orson Welles, sending much of New Jersey into a panic because it was taken rather too literally; in the cinema, with the special effects that came to characterise action films; with space travel, in pulp fiction and the fad for cult religions. Comical creatures from other planets formed the basis of children's TV series and entered a commercial relationship with the makers of dehydrated mashed potato.

Unidentified flying objects have been explained as the intergalactic transport of other worlds and, more confusingly, as relics from our own past. Thousands insist they have been abducted by aliens and interfered with, and there are people who claim not to be normal people at all but visitors from the planet Oobabado. In fact, all that is missing from our very well developed UFO and aliens culture is any concrete evidence of UFOs. Or aliens.

Except that if the conspiracy theorists are to be believed, aliens are already here, and have been for some time. In Roswell, New Mexico, in July 1947 apparently, and not far from where Goddard used to build his rockets, a farmer out on his ranch discovered some strange debris. This

debris has gone on to become the subject of much intense, sometimes frenzied, argument. The United States military maintain that what was found was nothing more interesting than the remains of a Project Mogul top-secret research balloon. But some still believe it was the wreckage of a crashed alien spacecraft, even going so far as to suggest that the remains of small alien bodies were found.

It is tempting to conclude that UFO sightings occur in areas such as Roswell, where something secret and military is usually going on anyway (around the time of the Roswell incident, the US Air Force was experimenting with numerous tailless flying wing prototypes). But Britain is not immune, and has recorded more than its fair share of UFO activity. Again, it must be acknowledged that a lot of this has centred around the West Country, where cider is popular and where the army engages in complex manoeuvres. It is also worth pointing out that Britain has more ghosts than any other country in the world. What is more, most UFO sightings stem from a time when the world was gripped by Cold War paranoia and trying to come to terms with the nuclear age.

Then again, the Ministry of Defence, which has consistently pooh-poohed the existence of flying saucers, has at times taken the subject more seriously than it might like to admit. The full extent of British interest in UFOs has only been revealed in the last few years. In the 1950s, the MOD organised a secret committee to draw up a report on a series of unexplained flying objects, to be presented to the prime minister, Winston Churchill.

At the time UFOs were making the headlines, and the notion that we might have been visited by aliens garnered support not just from mavericks, but also from well respected establishment figures such as Lord Louis Mountbatten. And it was in the *Sunday Dispatch* of 11 July 1954 that the notoriously dour Dowding wrote the words reprinted above.

Something called the Flying Saucer Working Party met in secrecy in a hotel near Trafalgar Square. It had just five members, and their job was to sift through hundreds of potential UFO sightings; after eight months, work, only a handful of cases were considered to have any merit.

One report came from an experienced test pilot. In August 1950 he saw a flat disc-shaped craft resembling, as he described it, 'a shirt button', light grey in colour, manoeuvring through a series of sharp turns at speeds estimated to be up to 1,000 miles per hour. It was flying over the Royal Aircraft Establishment at Farnborough. Then two weeks later the same pilot, along with five senior RAF officers, saw another disc-shaped object low in the sky, towards Guildford and Farnham. All six were questioned by an MOD team, and then advised not to discuss what they had seen. At first you would think this was pretty convincing stuff, but the Flying Saucer Working Party reached the conclusion that the test pilot's first sighting was an optical illusion; they said it was 'impossible to believe', that anything could have flown at high speed and low altitude, over such a densely populated area on a clear day without being reported by anyone else. As far as the second sighting was concerned, they felt the five witnesses must have been unduly influenced by the first report when they saw the flying saucer.

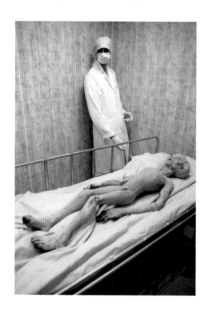

116—117

James May's 20th Century: We Know You're Out There. Don't We?

Giovanni Schiaparelli, the man who started it all.

Below:
A 1907 map by the American Percival Lowell, clearly showing signs of 'civilisation' on the red planet.

As a result of the investigation Churchill was told that, as far as UFOs were concerned, there was nothing to worry about. And, in reality, the government's real interest in UFOs had more to do with the fear of invading Russians than the dread of marauding Martians. In the days of the Cold War it was deemed vital to track any flying object that attempted to penetrate the UK's air defences, even if it was generally accepted that it had merely penetrated the darker recesses of someone's imagination.

Nevertheless, enthusiasm for flying saucers has never really gone away, not least because the prospect of meeting beings from another world is so compelling, even at the risk of extermination by a death ray accompanied by a strange buzzing noise made with an early synthesiser. The hope that there is something 'out there' can be traced back to the time when astronomers first turned their telescopes towards Mars.

In the late eighteenth century the astronomer William Herschel observed dark areas on Mars (Herschel was no fool. He produced the first credible map of our galaxy). He speculated these Martian dark areas were seas, full of water; this prompted the idea there may even be life on the Red Planet. In an address to the Royal Society he said: 'Mars has a considerable but modest atmosphere, so that its inhabitants probably enjoy a situation in many respects similar to our own.'

Then around a hundred years later, in 1877, another star-gazer, Giovanni Schiaparelli, made a more detailed study of Mars through a more powerful telescope. He noticed some curious features; there appeared to be lines on the surface of the planet. Schiaparelli described these unusual contours as *canali*. The word means 'channels', which should have been taken to be quite natural formations, but unfortunately *canali* was wrongly interpreted by some to mean 'canals'. Soon Mars was thought to be inhabited by a race of aliens who had the wit and wisdom to create a network of alien-made waterways. Although Schiaparelli favoured the idea that the canals were natural, he didn't oppose the idea that they might have been built. It has been suggested that this was to make sure he could get another research grant to pay for his star-gazing.

At the time the idea of canals and intelligent life didn't seem that absurd. After all, our earth is the third planet from the sun, and Mars is the fourth. It is only about one and a half times further away than we are, so it was more than possible it might have sufficient warmth for life. Also astronomers became intrigued by Mars's regularly changing polar caps. This occurs because the axis of Mars, like that of earth, is on a tilt; all of which supported the idea that the planet could have liquid water and seasons. The case for Mars as home to a life similar to ours was a strong one.

Speculation reached its peak at the turn of the century, when an American, Percival Lowell, began his observations of Mars from his observatory in Flagstaff, Arizona. He also saw what he thought was a complex system of straight canals crossing the surface of Mars. He, too, drew the conclusion that these lines were far too regular to be natural features. But Lowell went further. He suggested that perhaps the canals

'IT WOULD TAKE
ABOUT TWO AND
A HALF MILLION
YEARS FOR
A SIGNAL TO
REACH US FROM
A NEIGHBOURING
GALAXY... UNLESS
OF COURSE THESE
ALIENS KNOW
SOMETHING
WE DON'T'

were built to allow water to flow from the polar regions to the arid
equatorial regions. Soon imaginations were running wild, and Lowell had
populated Mars with an ancient civilisation of highly intelligent Martians,
trying desperately to survive on their dying planet by bringing water from
the poles to their great cities along the equator.

The public was entranced by Lowell's vision. Editors of respectable
newspapers, such as the *New York Times*, supported Lowell and criticised
other astronomers, who said they could only observe dark smudges instead
of lines, and who suggested that the canals were merely optical illusions,
distortions produced by observing the planet from tens of millions of miles
away (curiously, the *New York Times* was the very paper that later
dismissed Robert Goddard's rockets on the grounds that they wouldn't
work in the vacuum of space).

The idea of Mars as a dying planet was the spur to Wells's *War of the
Worlds* – what would this advanced but doomed race of beings do, if not

conquer the earth and comandeer all its resources? They get away with it, too, but only until they are all wiped out by Earth flu.

Subscribers to the life-on-Mars theory caught a bit of a cold as well. We now know for certain that there is no intelligent life on our nearest neighbour, and definitely no canals. Fortunately, just as Mars dropped off the alien radar, new technology arrived to help us look further afield for ET.

In the late 1950s a systematic search began. It was to be known as SETI – the 'Search for Extraterrestrial Intelligence'. The task employed the latest technology of the time, radio telescopes, in the hope of picking up signals from life on planets far, far away. They need not be signals intended for us; but if there was intelligent life out there then, the thinking goes, they should also have radio and television, or something quite like it. If so, their own version of *The Archers* would travel through the cosmos to be detected by us. It is worth saying that even travelling at the speed of light, or roughly 186,000 miles per second, these signals can take an awfully long time to get here. It will take a radio signal from the Alpha Centauri star system, the nearest outside our solar system, about four years to reach us. But to receive a signal from a galaxy that lies beyond our Milky Way means we will have to be very, very patient indeed. It would take about two and a half million years for a signal to reach us from a neighbouring galaxy such as Andromeda. Unless, of course, these aliens know something we don't.

In the 1960s, undaunted by the scale of the problem, the astronomer Frank Drake began searching for our neighbours. What is more, while he was waiting for the first messages from outer space to arrive, he filled his time by creating an equation that would help determine his chances of success. His equation attempts to estimate the number (N) of extraterrestrial civilisations that we might be able to contact. To keep things simple he confined his calculations to just our own galaxy, helpfully narrowing the search down from billions of billions of stars to just hundreds of billions of stars. It is known as the Drake Equation and this is it:

$$N = R^* \; f_p \; n_e \; f_l \; f_i \; f_c \; L.$$

As Professor Stephen Hawking's publisher once warned, every equation included in a book will halve its readership. Keep reading, though: Drake's Equation can also be used as a simple checklist of the different requirements that have to be met if we are ever going to hear from anyone even vaguely green and with an oversized head.

For example, R^* is the rate at which stars suitable to promote intelligent life are formed in our galaxy, and f_p stands for the fraction of those stars that have planets around them, just as in our solar system. After that, n_e is the number of planets circling each star that might be capable of sustaining life. In the case of our solar system it's just the one, earth.

Next in the line is f_l; this is the fraction of planets where life actually evolves: some scientists argue that where life can evolve it will – after all, it did here on earth. But others believe that life might be a very rare thing indeed, which is a bit worrying – but more of that later. Then f_i is the fraction of life that develops intelligence and here, once again, we hit uncertainty, because some scientists believe intelligence has such an

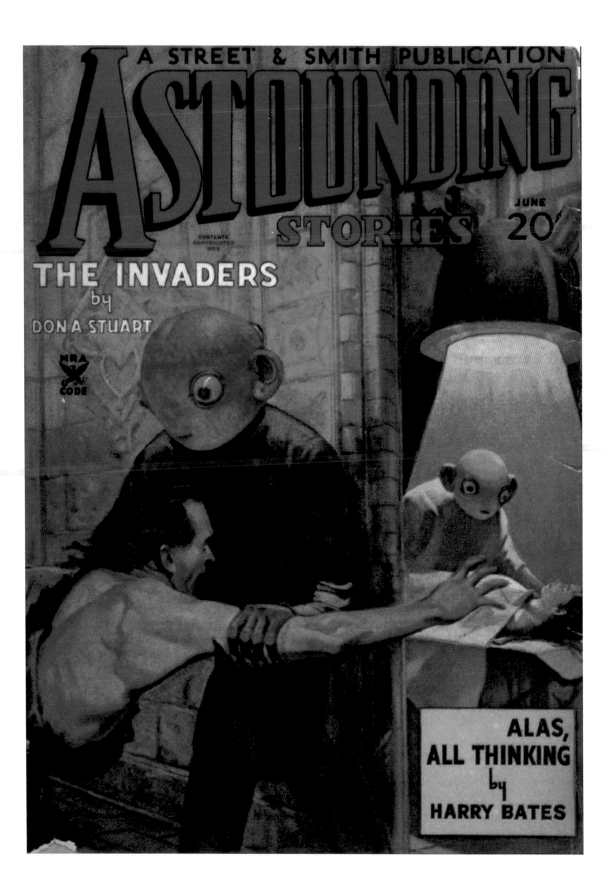

enormous survival advantage it will certainly evolve. However, others think intelligence is very rare.

That brings us to the final two factors: fc is the fraction of intelligent life that will communicate. Even if the aliens are intelligent, the question is, would they have the means and the desire to get in touch? Finally, L represents the time that the alien race survives and communicates. This is a really tricky question. If we take our planet earth, the lifetime of our sun and the earth is roughly 10 billion years. So far we have been communicating with radio waves for less than a hundred years. How long will our civilisation survive? Cosmologically speaking, not that long. And wouldn't that be annoying; to think that there were civilisations out there we could have spoken to, but which died out before we even said hello. And vice versa.

As you can see, we are not going to get a clear answer from the Drake Equation; its value lies in the questions it raises, rather than the answer itself.

SETI kept on searching for ET amongst the billions of stars, and in 1967 a signal did appear, though it was not discovered by SETI, but by Professor Antony Hewish and Jocelyn Bell at Cambridge University. To pick up a signal from space is not that unusual; many natural objects out there emit some sort of radio signal, our sun included. But what they spotted *was* unusual; a radio source with a regular pulsing signal.

It was Jocelyn Bell who noticed this signal. At first she thought it was an error, perhaps interference of some kind, but none the less, she took her findings to Professor Hewish, and they decided to see if they could track down the signal again. After searching for several months, they eventually picked up a clear series of rapid pulses, just a third of a second long.

It was possible that the source could be extraterrestrial intelligence. So for fun, Bell decided to name the radio source LGM–1, 'Little Green Men 1'. She hoped to keep LGM–1 secret until she had more information to work with. However, the news soon leaked out and Jocelyn Bell became something of a celebrity.

But what could the source of the signals be? Well, there were some clues. First of all the signal appeared to be from a stationary source. That meant it was unlikely to be a planet, as that should be in orbit around a star, and so would be moving towards us and then away from us. Eventually, the source turned out to be a star, spinning extremely fast, with a spot radiating a constant signal; the effect was the rapid switching of the signal, a sort of celestial radio lighthouse. It became known as a pulsar.

So it wasn't life, but all was not lost, because pulsars can be used as cosmic navigational points for probes searching the universe for new friends. The Pioneer 10 spacecraft, launched in 1972, carried a map showing the position of our solar system in relation to fourteen different pulsars. Using these, intelligent life should be able to work out where to find earth and plan a visit, or at least send an RSVP. Pioneer 11 followed one year later. Each craft carries a metal plaque showing the male and female human forms, the earth's position, and details of the crafts' journeys. They are expected to last longer than our own solar system.[4]

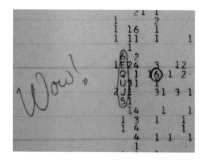

The so-called WOW! signal, detected in 1977 at Ohio University but not heard, or explained, since. Aliens, probably.

Bottom:
Traces of ancient bacterial life on Mars, recovered from a meteorite in Antarctica? Maybe.

Opposite:
A whole galaxy full of no alien life, perhaps.

Probes such as the Pioneers, and all those that have gone since, are a kind of reassurance; they give a sense that one day (whatever a day is out there) some other civilisation will find evidence of our own world, even if it turns out to be a posthumous message on the great galactic answering machine. At the same time, though, the need to send probes merely testifies to the growing futility of our search. In the Renaissance, it could be believed that there was life on the moon. Even less than a century ago, Martians were a credible idea. Now we know that we have no immediate neighbours in our solar system, or even far beyond that. The parameters of our search have expanded by an incomprehensible distance, and still there is nothing.[5]

In 1977 SETI did have one heart-stopping moment. The Big Ear radio telescope at Ohio State University detected what is now referred to as the 'Wow! signal'. It was a strong transmission lasting just seventy-two seconds, a jingle from outer space. But it has never been heard again, despite several frustrating searches.

Perhaps the closest we have been to identifying alien life was the result of a discovery not up there in space but down here on Earth. It crashed down onto our planet riding on ALH84001, a meteorite that hit in Antarctica about 13,000 years ago. In August 1996 ALH84001 caused quite a stir when it was announced that it might contain evidence of life. Scanning electron microscopes revealed strange structures on the rock, and if these tiny structures were fossilised bacteria-like life forms, then they would be the first solid evidence of the existence of extraterrestrial life.

The announcement of this potential alien life form caused considerable controversy at the time, and rekindled interest in Martian exploration. It even prompted the US President, Bill Clinton, to make an announcement on TV to mark the occasion.

However, most experts agree that the 'microfossils' are not indicative of life on Mars, but are more likely the result of contamination on earth. Even if they were unequivocally a record of bacterial life elsewhere in the cosmos, they didn't necessarily bode well for fans of Close Encounters, because in the late twentieth century a new theory came along to suggest that while very simple life forms might exist quite widely in the universe, the evolution of complex life, let alone the intelligent type, could be very unusual indeed.

The idea is called the Rare Earth hypothesis; it throws quite a lot of cold water on the chances of finding an extraterrestrial, at least one that we could talk to. The scientists who developed it argue that the conditions needed for the evolution of intelligent life are very, very rare. If their theory is right then the earth could be the only place in the Milky Way, and perhaps even the entire universe, where complex life exists.

For intelligent life to evolve (at least as we understand it), the right things have to happen in the right order. For a start it requires liquid water, which requires the sustaining planet to be the appropriate distance from the nearest sun. Too close and the water will boil, too far away and it will freeze. Of course a different atmosphere and different atmospheric

James May's 20th Century: We Know You're Out There. Don't We?

'BUT IS SPACE ACTUALLY INFINITE ANYWAY? WHO REALLY KNOWS? AND SHOULD WE WORRY? TO BE HONEST, THE CHAPTER ON CIVIL AVIATION IS OF MORE IMMEDIATE RELEVANCE'

pressure could have a bearing on all this, but that in itself would fundamentally alter the conditions for emerging life. The notion that the temperature has to be 'just right' is now known as the Goldilocks Theory.

The atmosphere of our hypothetical planet has to be able to maintain this temperature within fairly precise extremes. Planet Improbable also needs to be well away from X-ray and Gamma radiation, so will have to be far from neutron stars and quasars which produce both.

Ideally, it should also be close to a whopping great big planet such as Jupiter, as its powerful gravitational pull can attract all the asteroids and comets flying around the solar system and allow evolution to carry on without constant pummelling. A moon is also useful, in order to create ocean tides. Tidal activity creates tidal pools, which are great places for life to form. A magnetic field is also handy, as it protects a planet from solar wind and cosmic rays. And this is just an outline of the conditions that need to be in place for life as we know it, Jim.

On the other hand, some thinkers (me, to be honest, but a man from Jodrell Bank agrees) will point out that this view is slightly disingenuous, because we are looking back at a seemingly miraculous chain of events that resulted in the human condition as we know it, which is the only standpoint from which we can appreciate it. Other conditions might result in some other type of 'life', which we may not interpret as such. But now the

The alien visitors' car park at Roswell.

debate leaves the realm of science and enters the darkness of philosophy, which, historically, has driven many people to drink.

Either way, our being here on earth looks like being an enormous fluke, unlikely to recur anywhere else, despite the theory about monkeys, typewriters and Shakespeare that says in the infinity of space there must be another Earth somewhere; maybe an infinite number of them. (But is space actually infinite anyway? Who really knows? And should we worry? To be honest, the chapter on civil aviation is of more immediate relevance.)

But we'll keep up the cold calling anyway. In 1974 the Arecibo Radio Telescope in Puerto Rico sent a message into the cosmos. It consisted of 1,679 pulses beamed towards a cluster of stars called M13. If our signals are detected, and the recipients enjoy doing puzzles, they will work out that 1,679 is the product of the two prime numbers 23 and 73. And then if they arrange the pulses into a grid 23 by 73, they will discover a picture of a human being and a image of DNA.

But what if we are sending the wrong message entirely? The scientific community does seem terribly keen to bombard the universe with double biology and maths homework, and there is no reason to assume there will be any more enthusiasm for these things on Betelgeuse than there is on earth. How about a simple graphic representation of a pie and a pint, and some sort of universally understood cypher that says 'I'm paying'?

THE
IMPOSSIBLE CITY
BUILDINGS
ELECTRICITY
TRAFFIC

3

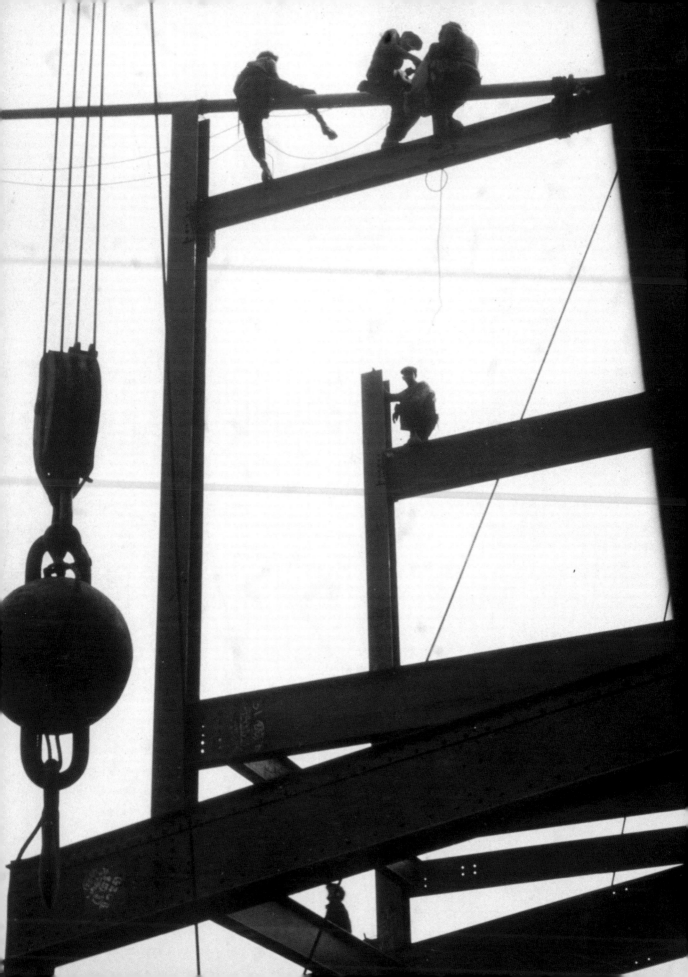

ROOM AT THE TOP

The steel frame of the Empire State Building under construction in 1930.

When the twentieth century drew to a close, it was still possible to enjoy a cold beer 107 floors up, overlooking Manhattan Island through the Windows on the World bar in the World Trade Center. It was one of the greatest unnatural landscapes on earth, the bar, the perfect vantage point from which to view countless glittering monuments to corporate triumph and vanity. From 1,300 feet up, Manhattan looked more like a trophy cabinet than a city.

On a clear day you could see for over forty miles. To the north were the towering Chrysler and Empire State Buildings. To the east, the incongruously ornate Woolworth Building, and beyond that the magnificent Brooklyn Bridge.

If you knew your geology you would be reminded that the powerful rivers that now surround Manhattan were once carved by huge glaciers that had bulldozed their way this far south during the most recent ice age. To this day the erratics, those enormous boulders in Central Park, remind us of this time. They were deposited 20,000 years ago when the glaciers retreated and the sea returned to form the island we recognise today.

Geology still holds Manhattan to ransom. Only the hard Manhattan schist provides the ideal bedrock to support skyscrapers. In Greenwich Village the soils are too weak to support the massive foundations needed to build to the sky. Space has always been at a premium on the island, and even the space that is there cannot be equally exploited. By the beginning of the twentieth century Manhattan was a popular one-way excursion destination for people from all over the globe who, after being checked over on Ellis Island for lice and intentions to overthrow the US government, were ferried across to the swelling metropolis to begin a better life. They all had to be housed and given a place of work, and as Manhattan is an island, too, any building expansion project was soon going to come up against the insuperable obstacle of the surrounding sea. So it was time to take the essentially medieval concept of the street, turn it through ninety degrees, and build upwards.

'THE GREAT PYRAMID OF CHEOPS HAS ONLY THREE VERY STUFFY BEDROOMS, NO GARAGE, AND NO WINDOWS'

Physics, however, throws up one of its inevitable objections at this point. For centuries the accepted way to build on any scale was with bricks or stones, but as anyone who has ridden a camel south of Cairo to visit the Pyramids of Giza will know, bricks and stones have their limitations.

The problem isn't actually one of height. The Great Pyramid of Cheops (or Khufu) towers 450 feet over the dusty plain and for 4,000 years was the tallest building in the world until pipped by Lincoln Cathedral. It contains some 2,300,000 blocks of stone with an average weight of 2.5 tons, although some weigh as much as eighty tons. It has been calculated that the volume of stone in the pyramid could build a wall six feet high all around France, which seems like a better use for it. It is reckoned to have taken 150,000 workers twenty years to construct, using techniques that remain something of a mystery. Yet for all this the Great Pyramid of Cheops has only three very stuffy bedrooms, no garage, and no windows.

There is no questioning the credentials of its designers – it is said to be built to within tolerances of 0.05 per cent – but there are problems with the materials. Put simply, stone is far too heavy, so as the height of the building increases, so too does the load on the lower floors – massively. In building a modern tower block it would be difficult to exceed ten storeys before the walls on the ground floor became unacceptably thick. Here, then, we arrive at an explanation of why the pyramid is the shape it is: not because it has esoteric powers or because the pharaohs thought it would aid the growth of their tomatoes, but because a 450-foot-high building made from stone must become progressively narrower if it is not to crush itself under its own weight.

The architects of the medieval cathedrals did not fare much better with stone. They had glass, which could be used to fill up large areas of the walls, but they still discovered (sometimes the hard way) that building upwards also necessitated building outwards, and that their cathedrals had to be almost as wide as they were high if they weren't to collapse. The returns for building high often diminished to the point where flying buttresses had to be supported by flying buttresses of their own (at York Minster, for example).

And so to France and a wonder of the modern world; more importantly, a structure that provides a clue to the way the builders of Manhattan would be able to break free of the bonds of Earth and its gravity. Gustave Eiffel's remarkable iron tower was the winning entry out of 700 submitted in a competition to design a new world wonder to commemorate the hundredth anniversary of the Revolution. It was erected, not without opposition, in 1889 (and was expected to last for only twenty years). The chattering classes, who felt it would destroy the ambience of their great city, called it the 'ugly skeleton', the 'giraffe cage' and 'an unprecedented crime against the art of architecture'. More worryingly, a professor of mathematics challenged Eiffel's calculations, claiming his tower would collapse when it reached 660 feet. It was only when Eiffel offered to guarantee the construction with his own personal fortune that he was permitted to build his tower, which rises to 985 feet and comprehensively trounces both Cheops's penthouse and Lincoln's parish church.

Gustave Eiffel (on the left) and an unknown friend at the foot of his 'ugly skeleton', the Eiffel Tower, in 1899.

Bottom:
The 'Flat Iron' Building, one of the first to use Eiffel's thinking in an inhabitable space.

Overleaf:
Steelworkers survey the miracle of modern Manhattan from a position 70 floors up on the unfinished RCA Building.

In fairness, the Eiffel Tower is not strictly a building, and has never found much favour as an office or apartment block. But if you stand at a distance and squint, it is possible to envisage it clad with walls and windows. Since the structure can stand up on its own, any cladding would have to fulfil no more onerous a function than weatherproofing the interior. *Voilà!*, as they never actually say in Paris.

Here, then, was the solution to Manhattan's housing and business premises problem. A skeleton of iron, or better still steel, would provide the structural integrity, and the gaps could be filled in with bricks and glass to make it inhabitable. Physics could be defeated and New Yorkers would for evermore only look upwards.[1]

This steel-frame technique was used for the first time in 1889 in Chicago, where George Fuller created the 165-foot Tacoma Building. That was demolished in 1929. In Manhattan, though, anyone strolling south down Broadway can see a surviving Fuller building. It is triangular in section and looks rather like the bow of a great ship, but the romantic locals refer to it as the Flat Iron. It was completed in 1902 and, fortuitously, a year later Elisha Otis perfected the electric lift. By balancing the elevator car and its passengers with a counter-weight, he was able to reduce the power required from the motor and the loads it would have to sustain. And by incorporating safety devices to arrest the car in the event of a cable failure, he was able to secure public confidence in the technology.[2]

The Flat Iron isn't huge, a mere 285 feet or twenty-two storeys, but it is traditionally considered the oldest surviving skyscraper in Manhattan. This honour should almost certainly have gone to the nearby Park Row Building, which is several years older and a hundred feet taller, but perhaps because its design has never found much favour with the locals, it is usually ignored.

Yet for all its beauty the Flat Iron didn't tower above the other buildings of the time in the way you expect, if not demand, of a skyscraper. It didn't even give the Great Pyramid a run for its money. But it wasn't to be long before a building came along that did. It was the new corporate headquarters of a global pick'n'mix empire: the Woolworth Building.

The Woolworth Building was completed in 1913 and was a step change in construction. It towers 792 feet above the Manhattan streets and boasts fifty-eight storeys. It was designed in a striking and rather un-Woolworthsy neo-Gothic style by architect Cass Gilbert and was commissioned by Frank Woolworth in 1910. As he might expect of his customers, Woolworth paid for the project in cash – a total of $13.5 million dollars in the money of the day. It was an object lesson in piling it high and finding it to be colossally expensive.

It was described as the Cathedral of Commerce by the reverend who performed the opening ceremony, a man who wouldn't have felt out of place in the vaulted and ornately mosaic lobby. There are even gargoyles, including one of Gilbert keeping a tight grip on his drawings and one of Woolworth counting out his dimes.

The Gothic stuff, however – indeed most of the building's solid areas – is something of a facade. At the time of writing the Woolworth Building is undergoing an overhaul, and on the forty-seventh floor the interior walls

The Woolworth Building.

have been stripped away to reveal the bones of the thing – the steel skeleton. It is little more than basic Meccano on a grand scale, and the fact that the floors above the forty-seventh are still in place is confirmation that the real work is being done by girders, not brickwork. The floors have no structural function at all.

The Woolworth Building was the tallest in the world – the Eiffel Tower, it was deemed, inspirational though it was, didn't count as it was not inhabitable. And French. The Woolworth Building held the record for almost thirty years, at which point it was overtaken by the 1,046-foot monolith that is the Chrysler Building, the first structure ever to surpass 1,000 feet. It was built very quickly, at an average rate of four floors per week, and, remarkably for the time, there were no fatalities among the workforce throughout the project. This was the heyday of architectural willy-waving, and during the Chrysler's construction a rival building at 40 Wall Street seemed intent on nosing ahead by a few feet. To break the unending mine's-bigger-than-yours cycle, Chrysler's architect, William Van Alen, obtained permission to add a 123-foot stainless-steel spire to the top of his challenger. Cleverly, though, he kept it hidden within the structure until the last minute, hoisting it into place just as it looked as though Wall Street had the biggest erection.

Size apart, the Chrysler Building is generally considered to be New York's best, a fabulous celebration of Art Deco and something of a brand-building exercise by the Chrysler Corporation. On the sixty-first floor can be found replicas of Chrysler bonnet ornaments, and on the thirty-first the floors are decorated in the style of Chrysler radiator caps. In the summer of 2005, a hundred architects, builders, engineers and historians, plus a few others selected from the city's great and good, were asked to choose their ten favourite high-rise buildings in New York. Almost all of them put the Chrysler Building in their top ten. When it was destroyed by heat-seeking missiles in the 1998 film *Godzilla*, cinema audiences were outraged.

Sadly for Chrysler, events would soon prove that 'tallest building in the world' is never an accolade that can be retained for long. The year after its completion, it became possible to look down on Chrysler from a bigger building, especially if you were an animated ape clinging to the top of a model. The Empire State Building rose to 1,250 feet (not including King Kong), boasted 6,500 windows and was served by seventy-three elevators. While it avoided the excesses of Woolworth's efforts, it still offers a few architectural highlights. The lobby – the very part that would end up as solid stone if traditional construction had been attempted – is an airy five storeys high. In the north corridor bemused visitors can see illuminated panels depicting the Empire State Building as the eighth wonder of the world, alongside the accepted seven. The Empire State Building would also generate better folklore, such as the case of lift operator Betty Lou Oliver, who plunged seventy-five floors in her car but survived. In 1979 Elvita Adams attempted to commit suicide by jumping from the eighty-sixth floor, only to be blown back in through a window one floor lower.

The rate at which it went up made Chrysler's project managers look

The anti-smoking regulations at New York's Chrysler Building were notoriously draconian.

positively tardy; 3,400 workers built it at an astonishing rate, sometimes faster than a floor a day. Their timing, however, was truly woeful. Its opening coincided almost exactly with the onset of the Great Depression in the US and, as a result, it was some time before much of its office space was occupied. It was dubbed the Empty State Building and did not become profitable for nearly twenty years.

The profitability issue was something of a wake-up call for the skyscraper. The Empire State was the pinnacle of its type, a very tightly woven framework of steel girders clad in stone. But architectural caution, doubts about the consistency of steel quality, and the frankly inadequate design techniques of the day led to massive over-engineering, as it had in the Woolworth Building. It is reckoned by modern analysts that the Empire State contains three times as much steel as it really needs to keep it up.

Over-engineering had two damaging side effects. The first was that all that steel greatly reduced the interior space available, and limited what could be done with it. In the Woolworth Building the biggest uncluttered room that could be incorporated was only as big as the distance between two uprights. The Empire State Building, despite once being the biggest in the world, is claustrophobic inside for similar reasons. The era of the communal open-plan workplace was still some way off.

The other problem was monetary. It is probably fair to say that without the indulgence of vain patrons or wealthy state departments, none of the early scrapers would have been financially viable. They were hideously expensive to build and, once up, did not yield as much rentable space as people might imagine. It was time to usher in the second generation of skyscraper, one in which grandiose detailing and cautious design gave way to easy-build austerity, precision engineering and acres of space that would generate bucks from the moment the epauletted doorman said welcome. It was time for the tower of power.

Curiously, one of the key elements of the new skyscrapers was the one material that had hindered the efforts of the early high-risers – glass. At the

turn of the century glass was a problem. It was a low-tech material, usually full of tiny flaws and hence extremely brittle. Everyday domestic window glass is still the same, which is why glazing is such a secure business.

Making really big panes in normal glass was difficult, and even when one was made it would be incredibly vulnerable at the top of a skyscraper, where wind pressure alone could be enough to break it. And repairing windows 1,000 feet up is an irksome business. An alternative was plate glass, which was cast on an iron surface but had to be ground and polished to a smooth finish, which was hideously expensive.

But in 1953 Alastair Pilkington and Kenneth Bickerstaff developed a new process in which the molten glass is poured onto the surface of molten tin to produce what is known as float glass. They did not strictly invent this, as something similar had been described in earlier US patents, but, in the language of industry, they 'productionised' the method to enable huge panes to be made in huge quantities.

Float glass is free of the flaws which, in conventional glass, concentrate stresses and weaken it. It requires no polishing, because the molten tin produces a perfect surface finish. Most importantly, controlled cooling of the glass toughens it to the point where a sheet barely twice as thick as a normal window pane can be vigorously belted with a hammer to no avail. It's a wonder beer glasses aren't made of this stuff.[3]

Now glass could become a building material in its own right, rather than just something for filling in holes and letting in light. Why put a window in the wall when you could make the whole wall out of a window?

At the same time the basics of skyscraper engineering underwent a radical rethink. Instead of the complex geometric steel skeleton, the new scrapers would be built around a technique sometimes known as the tube-within-a-tube method. A central solid core houses the lifts and other services. Around this is an outer shell, often made of little more than steel supports and the new glass. In between are massive and uncluttered decks, ideal for space-efficient open-plan offices or shopping malls.

Also crucial to the success of the phase two skyscraper were better quality steel and more precise engineering methods. More complete analysis of the loads on structures allowed architects to put their steel frames only where they were needed, and to use only enough material to withstand the loads they sustained and the worst-case effects of wind and possible earthquake. The buildings could be just strong enough. Size for size, this new generation of skyscraper could be built more quickly, using less material bulk, and would yield more rentable space.

In New York, the tube-within-a-tube skyscraper was epitomised by the 1,731-foot World Trade Center, opened in 1973, the Twin Towers of which were often described by the locals as looking like the boxes the Empire State and Chrysler Buildings came in. They were elegant and simple structures of glass and steel, and a towering testimony to the efficacy of the new building techniques. The steel used in the Towers' outer core was engineered so precisely that its cross-section reduced in thickness as the building rose higher to lighten the load. Commercial estate agents may like

James May's Twentieth Century: Room at the Top

to know that the World Trade Center boasted 13.4 million square feet of usable space and housed 50,000 rent-paying employees.

The building also employed a novel elevator arrangement that had only been tried once before, that of 'sky lobbies'. Fast express elevators travelled the full height of the Towers but stopped only at selected floors. From these, travellers could take a shuttle elevator that stopped at all floors in one section of the building. To reach a particular floor involved changing from a fast lift to a local one, as on a cross-country railway journey.

After a vertiginous beer in the top-floor bar, many visitors would take the trip to Staten Island for the simple pleasure of looking back at the tip of Manhattan. The impression of incredible wealth and commercial omnipotence was similar to that sensed when approaching Venice from the south with St Mark's Square in full view.

When the Twin Towers came down on 11 September 2001, some proclaimed that the event symbolised the collapse of capitalism and was a warning against the conceit of globalisation. Ancient texts in which the destruction of great towers was prophesied were quoted, as if it had been known all along that building so high was folly.

In fact the Twin Towers had already done their vital work in setting an example to the world. The same basic building techniques, bolstered by even more thorough structural analysis using powerful computers, have been used to build seemingly impossible towers all over the planet, and skyscraper design has become something of an off-the-peg science. In Manhattan, it was necessary to build up to save space. Now it is accepted that tall and incredible buildings are great PR and great morale-boosters, and that there is no better way to celebrate modernity than with a tower reaching into a region that, not much more than a hundred years ago, was regarded as the preserve of God and the birds. Even Dubai builds upwards, and it has a whole desert to expand into.

Soon a new skyscraper will touch the clouds where the Twin Towers stood. Let's hope that the architect and the people running it have the confidence to put another bar at the top, so that people can say, 'I'll have a tall one, too.'

The Manhattan skyline before
11 September 2001.

ONE NATION, ONE VOLTAGE, ONE PLUG

'THERE ARE PEOPLE ALIVE TODAY WHO CAN REMEMBER LIVING WITHOUT THE ELECTRIC LIGHT'

If you are ever in Hong Kong with a hour to spare in the evening, take a ride on the Star Ferry to Kowloon and join the many other tourists along the Tsim Sha Tsui Promenade, which looks south across Victoria Harbour to Hong Kong Island.

At night it is a mesmerising sight. Many skyscrapers compete for your attention. Like I.M. Pei's extraordinary Bank of China, extraordinary not just for its beauty but for daring to ignore the principles of feng shui. Or there's Building Two of the International Finance Centre. It's like a modern day reincarnation of the Empire State Building and, at 1,364 feet, very similar in height. But it is far more elegant.

As the time approaches eight o'clock, the crowds along the waterfront become still. They are about to witness one of the most unusual public entertainments in the world. Exactly on the hour music bursts out from huge hidden speakers, and the distant waterfront comes alive. Suddenly the tall buildings are no longer the headquarters of merchant banks, or the nerve centres of multinationals. For the next few minutes they reveal another side of their personality and dance with light, because the skyscrapers have all been wired up to perform a *son et lumière* of gigantic proportions. Tungsten and fluorescent lights chase up the sides of some of the world's tallest buildings. Neon dazzles with impossible colours and laser light cuts into the sky. It's all terribly kitsch, and looks like a scene from Walt Disney's *Sorcerer's Apprentice*, but you have to admire the imagination and ambition of whoever came up with the idea of painting with light on such a grand scale. When the show ends, the office buildings get back to work and their well-paid executives return to their negotiations as if it never happened.

The Hong Kong light spectacular is ostensibly a bit of entertainment, little more than showing off. At the same time, though, it is something of a paean to the glorious usefulness of a universal and dependable electricity supply and to the life-enhancing properties of artificial light. Remarkably, these things are relative newcomers to the city. There are people alive

The Hong Kong Harbour light show.

Even in 1919 they were usng sex to sell light bulbs.

today who can remember living without what very old people still sometimes refer to as 'the electric light'.

Light, especially, has always been coveted. In religious and folkloric imagery light represents truth and salvation, while the darkness is host to evil and death. The psychological trauma of living in constant gloom – for example in cities above the Arctic Circle during winter – is well documented. In the middle ages, light was so important to the scholarly monks that part of their wages was paid in candles. Without a good stock of candles they would be plunged into darkness as soon as night fell and work on the Lindisfarne Gospels would have to wait until the next day. This manuscript was the work of a monk called Eadfrith in the late eighth century. He sketched the layout on the reverse of each folio and then lit it from behind to produce a sort of illuminated-manuscript-by-numbers. The effort appears to have killed him and the work is not entirely finished.

Even at the start of the twentieth century, cities spent much of the time in darkness. To be fair, there was gaslight, but it was a dim substitute for the real thing and didn't extend far. It provided the perfect backdrop for villainy of all varieties, so much so that wise West Midlands scientists back in the 1770s chose to gather on the full Moon so they could readily find their way home after dark, and so became known as the Birmingham Lunar Society.

The wait for electric light was a long one. The basic principles had been around since the early 1800s; in fact the early electric light predated the gas mantle, its ultimate competitor, by a decade. But the early lights were carbon arc types, in which an electrical spark was made to jump across a gap between carbon electrodes, as first demonstrated in the early 1800s by Humphrey Davy of miners' lamp fame. The carbon arc method is still used in, for example, some searchlights. But they were expensive and difficult to maintain, as the electrodes wore out and required constant adjustment. So they were suitable only for a few well-maintained public places.

A better solution was the incandescent light bulb, of the type we now know, but getting one to work – at least for any length of time – proved tricky. The filament tended to burn out. The first person to achieve any real success with the light bulb was a Newcastle scientist, Joseph Swan, who in 1879 demonstrated that a filament in a vacuum could be made to glow. Swan quickly filed a patent to that effect. He also patented bromide paper for producing black-and-white photographic prints. It's still used.

But it was the American Thomas Edison who made the bulb a commercial proposition. A year later, after endless experimentation with different materials and different gasses for the globe, he managed to make a bulb with a filament of carbonised bamboo that would glow for 1,200 hours. (In 1991 Phillips invented a bulb based on magnetic induction that lasted for 60,000 hours.) Not only did Edison's bulbs work, they were cheap to produce, and so he soon began making tens of thousands of them. They were still not perfect, because decay of the filament would gradually produce dark spots on the inside of the glass globe. In 1903 Willis Whitney designed a bulb with a metal-coated carbon filament, the forerunner of the modern tungsten bulb, to solve the problem.

So towards the end of the nineteenth century, the light bulb had become a reality. In 1880 Swan's house in Gateshead became the first in the world to be lit by electricity. The year before, Edison had illuminated his Menlo Park laboratory in New Jersey using electric light bulbs. Swan lit up the Houses of Parliament in 1881 and the British Museum in 1882. But these were mere stunts in the great scheme of things, because lighting up a city requires much more than just a bulb.

Edison responded to the new craze for electric lighting by building a public power station on Pearl Street in New York's Lower Manhattan. It wasn't the first public electricity utility in the world – a distinction claimed by the water-wheel-driven generator in the Surrey town of Godalming – but it was a genuine attempt to put power not just on to the streets but also into the home. At first, electricity was seen simply as a means of lighting, and so the station, which was driven by a steam engine, ran only at night. Soon, though, the demand from industries using electric motors and the enthusiasm for new domestic electrical appliances led to constant working. The era of constant, public, on-demand and billed electricity had arrived.

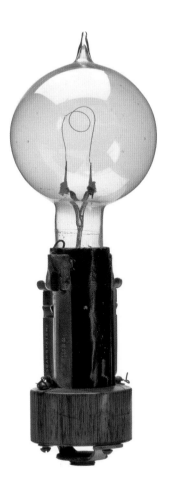

An early Swan light bulb. Creating a near vacuum inside the globe was the key to a long-lasting filament.

Overleaf:
A satellite map of the world at night. All that light is coming from bulbs of some sort.

Even in the 1900s, sex was being used to sell portable table-top electric fans.

Or had it? As the cult of electricity took hold of the developed world, a few basic problems emerged. The first was that electricity generation was still a very local affair, and different generating companies settled on different and seemingly arbitrary voltages. This made life difficult for the makers of electrical appliances. A standard electric bed-warmer could not be made for the whole world, or for a whole country, or even for a whole city. So a travel iron was unthinkable.

Another issue was the nature of the electricity itself. Some pioneers, such as Edison's great rival supplier George Westinghouse, advocated the use of alternating current (AC), of the sort that springs naturally from a spinning generator and in which the polarity of the supply oscillates at high speed. Others, Edison amongst them, championed direct current (DC), which can be rectified from an AC generator to give the sort of supply achieved with a battery. The electricity supplied by the spinning alternator of a car, for example, is AC, but it is rectified before feeding the DC electrical systems of the car and recharging the battery.

There were arguments for and against both. DC was easier to manage and simplified the design of appliances. On the downside, it was difficult to distribute over any distance because of voltage drop in the supply cables, which were required to grow fatter the farther they travelled from the generating station and used a great deal of expensive (and scarce) copper. Transmitting DC over a distance of much more than a mile was impractical.

AC was ideal for distribution over long distances. The Serbian-American Nikola Tesla demonstrated that relatively simple transformers could be used to 'step down' a very high supply voltage to a more usable one for the home or factory. High supply voltage meant that widescale distribution could be achieved with relatively light cables and with minimal losses. The physics behind all this is surprisingly complicated. And boring. So we'll leave it there.

However, AC was generally considered more dangerous, not least because its constant oscillation could confuse the heart muscles and cause ventricular fibrillation. This last point is deeply significant, and is central to one of the most distasteful public relations conflicts in history, the so-called 'war of the currents'.

Essentially, Edison, who had already invested heavily in DC, began a smear campaign against Westinghouse and his AC system. In an attempt to demonstrate the lethal flaw in the AC approach, he connected metal plates to an AC generator and lured stray cats and dogs on to them to suffer electrocution, a term which seems to have been coined at exactly this time. Members of the press were invited to witness and photograph these experiments. The whole business culminated in the destruction of Topsy the elephant. This may explain why Edison's reputation is rather tarnished in Britain especially. He may have been the creator of the record player, but he wasn't very nice to animals.

Now the campaign took a particularly macabre twist. In 1886 New York State arranged a review board to consider methods of capital punishment to replace hanging. Death by electrocution was proposed and accepted, but it remained to decide upon the type of current used. Edison hired the inventor Harold Brown and his assistant Dr Fred Peterson to design an electric chair, and to perform demonstrations showing that animal subjects suffered horribly when it was wired up to the 'safe' DC but met a swift end under AC. Edison reasoned that no-one would want the current of judicial execution in the home.

Since Dr Peterson sat on the review board, it comes as no surprise to learn that AC was adopted for Old Sparky. (Edison even attempted, with some success, to turn 'Westinghouse' into a verb. Some people referred to death by electric chair as being 'Westinghoused'.) Westinghouse objected to the board's decision and refused to supply the electrical equipment required by the prisons; Edison simply renounced DC just long enough to do so himself. Ultimately, though, he lost the war of the currents, because practical experience soon showed AC to be the way ahead. Oddly, Edison had once declared himself against capital punishment. He is also said to have admitted privately, late in his life, that he knew all along that AC was probably better.

Fortunately dragging Britain out of the dark ages was not to be so gruesome, even if it was somewhat protracted. Here one man saw the electric light and saw that it could be good. He was Charles Merz.

Merz was an advocate of large-scale electricity generation using AC and the new Parsons steam turbines. His Carville station on Tyneside, opened in 1904, was a model for the future and supplied an area unprecedented in the story of electricity networks. His ambition was to supply the whole country.

Just as in the US, electricity supply in Britain in the early 1900s was a joke. Generators, many of them DC, had been knocked up as and when required. As late as 1960, my great grandmother's rural West Country house was still supplied with local DC, and my father spent some time tracking down a replacement DC cooker for her. By 1916 there were over 500 different electicity suppliers and a bewildering array of voltages in use. Even in 1922 there were over twenty domestic socket types. The future of Britain was put on hold while everyone scrabbled around looking for a suitable plug.

Merz was convinced that what the cities needed – what the whole country needed – was a standardised, co-ordinated, universal electricity supply. And eventually, in 1926, after two decades of cajoling and coercing, and going to many boring meetings, Merz finally won his argument. An act was passed in parliament to create the National Grid (or 'Gridiron', as it was first known).

There was one more hurdle to consider before the people of Britain could embrace the travel hair dryer and the trouser press – the small matter of some pylons. Around 30,000 were required for the first phase of the grid, and the government held a competition to see who could design the best looking one. The winners were the Milliken brothers, whose design is the one we still see today. One wonders what the losing entries looked like.

But this new national plan nearly ruined its creator. Unfortunately for Mertz, the new standardised voltage that was adopted was not the same as the one he was supplying from his own power stations in the north-east. Fortunately for Merz, the government was so pleased with his work that it diverted a substantial amount of cash from the unemployment benefit fund to save him from bankruptcy.

And why not? Mertz is a true forefather of the modern city, a man who ensured that a predictable electricity supply could be taken for granted in the way air to breathe can be. Electricity had been around long before he realised his vision, but until he standardised it, the city floundered, stumbling in the twilight of the Victorian era and unable to embrace the high-rise, twenty-four-hour, permanently lit ethos that has since come to define it. Electricity, perhaps, is the opium of the people.

Consider the impact of a power cut in the evening on a winter's day. An inconvenience, certainly, but so much more than that: a stark reminder of how grim city life must have been for the vast majority of people, not much more than a lifetime ago.

A modern pylon, still essentially to the original Milliken Brothers' design.

REGAINING CONTROL OF THE CAR

It was a while before town planners realised that traffic lights should go on poles.

The emancipation of the horse was a long time coming. The advent of steam power in the industrial revolution was the first blow for its freedom, since steam would inherit the yoke the horse had worn for thousands of years in service to humankind's quest for productivity. James Watt, the inventor of the separate-condenser steam engine, quantified a unit of horse power for his machines simply because that's how his customers would specify them: in terms of the number of horses they would replace.

The beam engine liberated the horse from its duties in the pumping houses of pits and in small factories everywhere. Later steam engines would banish horses from the towpaths of canals. The coming of the railways spelt the end of the stagecoach and down on the farm the traction engine gradually replaced the plough horse.

But it was not until the twentieth century that the horse was released from the worst of its bondage, which was the role it played in our towns and cities. The romantic vision of the horse in Victorian Britain is of an elegant thoroughbred trotting along at the head of a light trap occupied by a genteel lady. In reality, most city horses were heavyweight beasts of burden living a life so gruelling that even a healthy one might live for only two or three years before dying of exhaustion. Final and merciful release for the horse would come only with the invention of the internal combustion engine and the motor vehicle.

The growth in vehicle ownership during the twentieth century was phenomenal, and within one generation the whole culture of horse-drawn transport had been put out to grass. This was generally for the better, and not just for the horse.

Many of the problems we associate with the motoring age were already troubling people in the horse era. In London, in the late nineteenth century, some 300,000 working horses took to the roads each day. There were horse traffic jams, horse accidents, horse delays, and probably horse-generated road rage. Horses were expensive to run and horses were stolen.

1905, and there is still some confusion as to the best form of propulsion. But not for long.

Horse emissions were pretty bad, too. A typical large horse produces around nine tons of manure a year, and perhaps 1,500 pints of urine. So it is easy to see why some Victorian statisticians prophesied the eventual submersion of whole cities under horse pooh. And while horse manure in small quantities is unlikely to infect humans, in large piles it can spread E Coli. More ominously, horse manure is a fertile breeding ground for flies, and studies have suggested that the disease spread by these flies was the primary cause of the terrible infant mortality rate in nineteenth-century cities.

So the coming of the motor vehicle should have been seen as a good thing. Even so, there were many – especially those in the pro-horse lobby, who clearly believed that this noble beast was not yet done with suffering – who wanted the new machine to be tamed.

The most draconian piece of anti-car legislation ever written, Britain's infamous Red Flag Law, had already been repealed in 1896. This had required all motor vehicles to be preceded by a pedestrian waving a red flag or a lantern, to warn people of the approaching 'danger'. Speeds were limited to 4 mph in the country and 2 mph in towns, where cars were always required to give way to horses. The London to Brighton veteran car rally was inaugurated to celebrate the demise of the red flag, and is still run today, lest we should ever forget that cruelty to cars was once officially condoned.

With the flag-waving over, it made sense to develop some more pro-active ways to ease the relationship between the metropolis and the car,

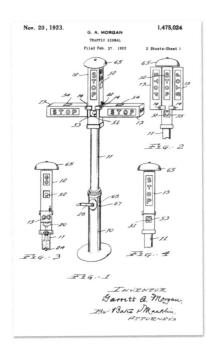

Nov. 20, 1923.   G. A. MORGAN   1,475,024
                    TRAFFIC SIGNAL
                   Filed Feb. 27. 1922   2 Sheets-Sheet 1

which was likely to be an uneasy one nevertheless. One vital component of modern traffic management had actually been in place in the horse age. In 1868, what we would now call a traffic light was erected in the road outside the Houses of Parliament, where congestion was notoriously bad. It featured red and green lanterns and was manually operated by an attendant policeman. Unfortunately, this being a time before mains electricity, it was gas-powered and rather predictably exploded (the policeman was injured, but lived).

It would be a while before traffic lights as we know them were created, but a number of ingenious systems were tried, including a clock-like device with a rotating hand that pointed alternately to red and green quadrants of its face. These things were sometimes referred to as 'robots'.

The first patent for a traffic signal was granted in the US, in 1923, to Garrett Augustus Morgan of Kentucky, the son of a slave family. He had little in the way of formal education, but presumably had a shed of some sort, and enjoyed experimenting with gadgets while earning a crust repairing sewing machines. He seemed to have a penchant for safety equipment. His first invention was the Morgan Safety Hood and Smoke Protector, which earned him national fame in 1916 when it proved crucial in the rescue of thirty-two men trapped in a tunnel 250 feet below Lake Erie.

The inspiration for the traffic signal was an accident he witnessed between a car and a horse-drawn carriage, in which a young girl was badly injured. His semaphore system was more like the signalling used on the railways, with a rigid flag atop a pole that could be set to one of three positions: 'stop', 'go', and 'everything stop', the latter to allow pedestrians to cross safely. It was tested in Cleveland in 1914 and was so popular that Morgan eventually sold the patent rights to the device to General Electric for $40,000.

Top:
Morgan's original patent application for a 'traffic signal'.

Right:
The ceremonial unveiling of a new traffic light on Fifth Avenue, New York, in an era when people still thought them a good thing.

Inserting catseyes in the middle of the road required skill, bravery and minimum traffic. A good flat cap was useful but not essential.

Today's coloured light signals retain the basic logistical principles of Morgan's semaphore, and the convention for red and green lights meaning 'stop' and 'go' respectively has been adopted by most of the world. Exceptions are Spain, southern Italy and India (Britain's first automatic signals were installed in Princes Square, Wolverhampton, in 1927).

The roundabout, too, appears to be in essence an American invention. The first 'traffic circle' was built in New York in 1905 and was designed to allow joining traffic to keep moving. Later traffic circles (or 'rotaries') employed signals to stop circulating traffic so that approaching vehicles could join unhindered. Unfortunately, this system tended to cause congestion on the traffic circle itself – which in the US is usually relatively large – and the feature fell out of favour in the 1950s. The shortcomings of giving priority to merging traffic are still apparent at the Arc de Triomphe in Paris.

Britain, though, persevered with the generally smaller 'roundabout', and in the 1960s established a ruling that said joining traffic must give way to that already in circulation. The efficacy of this is now recognised everywhere except rural France (this is surprising because it was the French architect Eugène Henard who was devising circular systems of traffic management as early as 1877).

Britain also gave the world – although not much of it, because it has not found much favour elsewhere – the catseye. It was the creation of Percy Shaw, a road-builder from Halifax, and patented in 1935. The role of the real cat in the invention is confirmed by surviving members of the Shaw family. Percy Shaw would often drive late at night from his regular pub to his home in Boothtown, and in the inevitable fog would navigate by the reflection of his feeble headlights from the polished tramlines set in the road. When the tramlines were removed, he had difficulty finding his way and one night stopped to investigate two pinpricks of light he saw in the road. He realised that they must have been the eyes of a cat; realised, also, that if he hadn't stopped at that moment he would have driven over a steep precipice.

A type of reflective stud, similar to the ones Shaw would eventually use, already featured on some road signs, so he simply nicked some and worked out his device. His genius became apparent in the Second World War , when catseyes were found to be clearly visible in the shrouded beams of blacked-out vehicle headlights but without reflecting light into the sky, where it might attract enemy aircraft. They were, and still are, made in such a way that when a car runs over the rubber shroud holding the 'eyes', it is compressed to be wiped clean by a rubber strip. An incidental benefit of catseyes is that they act as a rumble strip to alert drivers who have nodded off at the wheel.

The familiar Tarmac road surface is also the outcome of a moment of serendipity. Roads had been macadamised since the nineteenth century by rolling layers of progressively smaller stones, and eventually gravel, on top of one another. But macadamised roads, though a great improvement on cobbles, were not fixed, and so soon became rutted by wheels. In 1901, Edgar Hooley, a surveyor for Nottinghamshire, was inspecting a newly finished stretch of road when he noticed that one small section was

'THE ROLE OF THE REAL CAT IN THE INVENTION IS CONFIRMED BY SURVIVING MEMBERS OF THE SHAW FAMILY'

especially firm. It was explained that a barrel of tar had been spilled, and that road builders had spread slag on top of it and rolled it flat in an attempt to make good the damage. The combination of the macadam technique and the tar sealant became known as tar macadam, or simply Tarmac (the British company called Tarmac is a direct descendant of the one Hooley formed).

But many of the innovations that made motoring safer would become part of the vehicle itself, and in the early 1900s they came thick and fast. The klaxon horn first honked in 1901. The speedometer was introduced by the British company Thorpe and Salter in 1902, and was calibrated up to a hair-raising 35 mph, even though the speed limit at the time was 20 mph and wasn't to change until 1930. Windscreen wipers were conceived in 1905, when Mary Anderson of Alabama was watching her driver struggle to keep snow off the windscreen. She suggested a wiper operated by a lever inside the car. Even after electric motors were used to power windscreen wipers, many cars retained the manual lever for use in an emergency (up until the Second World War, the motor was often an optional extra in more basic cars). The rear-view mirror was patented in France in 1906, but has never been used there.

Parking meter misery came to Britain in 1958. Here, a milkman is robbed in Mayfair.

All of this helped fuel the huge increase in vehicle ownership and use but, inevitably, revealed a downside to motoring; namely, that all vehicles are a nuisance when they are not actually being driven. Early efforts at what is now rather confusingly called 'parking enforcement' (since its aim is generally non-parking) were made in the US, where policemen were empowered to mark the tyres of parked cars with chalk and then fine their owners if they were shown to have been parked for too long. Unfortunately, the system was thought to take up too much valuable police time, and it was generally accepted that any reasonably imaginative motorist could simply rub out the chalk mark and feign innocence.

As a result of this, in 1932, a newspaper editor from Oklahoma called Carl Magee came up with idea of a parking meter. The first man to be fined for overstaying his metered time was a man of God, who returned to his vehicle a month late. In Britain we were spared the tyranny of the parking meter until 1958, when the first ones were installed in Mayfair.

From the 1960s onwards, ideas for managing traffic proliferated. Elaborate one-way systems became very fashionable with town planners, but are now believed to facilitate reckless driving. Bus lanes, pedestrianised town centres and park-and-ride schemes have also blotted the townscape. Roundabouts, once intended to smooth the progress of vehicles, have become the means by which local authorities deliberately frustrate traffic flow in the interests of 'calming', and join speed humps and width restrictions in inflaming the tempers of drivers.

Put perhaps the greatest abuse of a fine innovation designed to aid the motorist and pedestrian alike is the indiscriminate use of traffic lights as a universal panacea to every perceived traffic problem. Their overuse led to some interesting experiments during the 1990s, during which all traffic controls in designated areas of large cities were turned off so that the effect on traffic flow could be observed.

It flowed better.

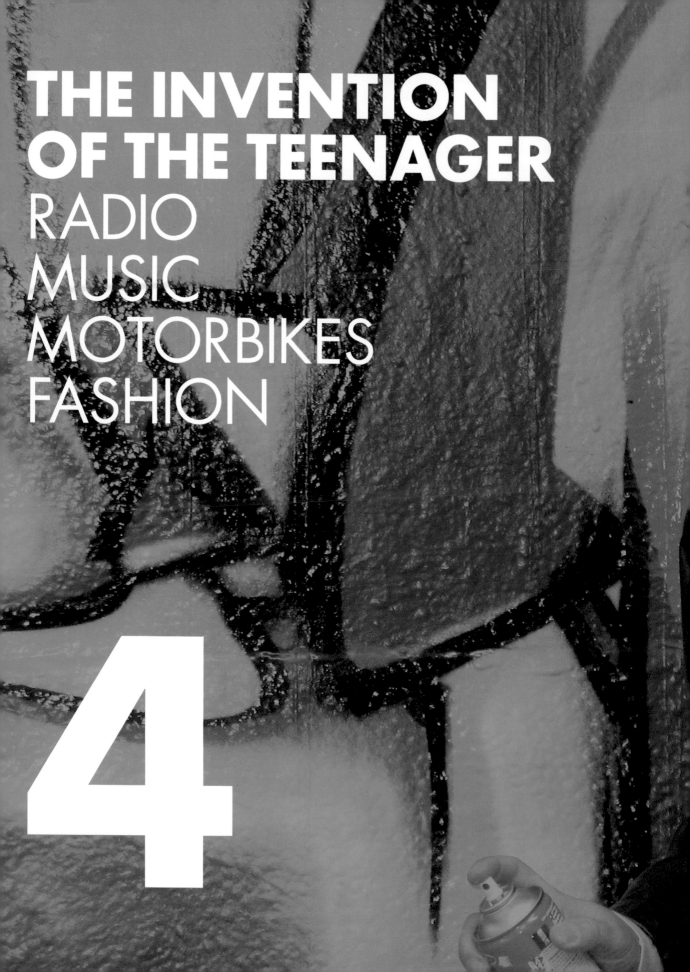

THE INVENTION OF THE TEENAGER

RADIO
MUSIC
MOTORBIKES
FASHION

4

MAKING WAVES

I'd sit alone and watch your light
My only friend through teenage nights
And everything I had to know
I heard it on my radio.
Queen, 'Radio Ga-Ga' (1984)

At the dawn of the twentieth century the teenage bedroom was a pretty dismal place. Largely empty, probably finished in predominantly brown hues, soulless, often cold, it was a monument to how quickly a sixteen-year-old had grown up.

Its occupant might well already be in full-time employment: in domestic service, as a clerk, a farm hand or a factory worker. There may have been a picture or two on the walls and some books and magazines on a shelf, but of the accepted trappings of adolescence – the pop posters and discarded trainers, for example – there would have been none. Most significantly, the teenage lair of 1900 was a place without music.

The phonograph would help; the vinyl single and Dansette record player would help more. But teenage salvation of the sort Queen celebrated would have to wait for the invention of radio and, more importantly, a low-priced portable radio on which there might be something worth listening to. And that was still half a century away.

The invention of radio, like that of television, is impossible to ascribe to a single person. First, the physical phenomena that would lead to radio would have to be discovered. Then they would have to be applied to something usable. Even then, the concept of 'broadcasting', as we know it, had to be envisioned and the culture of the radio station established; and further technological breakthroughs were required before old people were in a position to worry about rampaging youths spending their whole lives with small, brightly coloured plastic boxes clamped to their ears. But it started with a spark.

Harlem, New York 1984: a teenager shows how to wear a Boom Box.

**'JAGDISH
CHANDRA BOSE
USED RADIO
WAVES TO RING
A BELL AND FIRE
A CANNON
REMOTELY'**

That spark was generated by the German physicist Heinrich Hertz and his 'magic ring'. In a classic experiment, conducted in 1888 in Bonn, he produced a radio wave at one end of his laboratory and received it at the other. This wave was produced with a spark transmitter, in which a high voltage (several thousand volts, in fact) caused a small spark to jump across a gap. His 'receiver' was a metal ring featuring a similar small gap. When a spark was made at the transmitter, a second one was seen in the small gap of the 'magic ring', thus showing that the radio wave had passed through the air.

It wasn't much of a transmission, and Hertz told his pupils that he saw no useful purpose for his work, but that spark was enough to energise a frenzy of research into the practical possibilities of radio. Two years later, the Frenchman Edouard Branly discovered that the electrical resistance of a tube of iron filings reduced dramatically when a spark was discharged nearby, thus providing a reliable means of detecting electromagnetic activity.

In 1893 in the US, Nikola Tesla, a genius who was also to work in the fields of robotics, computers, ballistics and nuclear science, laid the foundations of modern radio. Next, the Russian Alexander Popov transmitted radio signals between the buildings of St Petersburg university and became – certainly as far as the Russians are concerned, and in learned circles, too – the inventor of radio.

Meanwhile, in India, Jagdish Chandra Bose used radio waves to ring a bell and fire a cannon remotely, thus laying the foundations of radio control if not Radio Luxembourg.

But the name most famously associated with radio, and to which its invention is usually attributed, is that of Guglielmo Marconi, the Bologna-born son of an Italian landowner and an Irish mother. At his death in 1937 radio stations the world over fell silent for two minutes to confirm his status as the father of the medium.

Marconi in his radio room, listening to the radio.

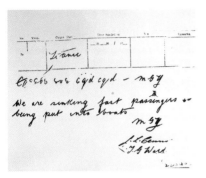

The last radio message received from the *Titanic*, 14 April 1912. This SOS was crucial to the rescue of over 700 survivors.

Marconi was utterly obsessed by radio waves from an early age. By 1894 he was transmitting signals over a hundred yards in the garden of the family estate. A year later he was achieving distances of over a mile. He had difficulty, however, in transmitting his enthusiasm to the Italian Ministry of Posts and Telegraphs, who were completely uninterested in his work. So in 1896, after a fortuitous introduction to the far more receptive Engineer in Chief of the Post Office, he moved to England.

At this point it is worth taking a brief pause in the story of radio to deal with some definitions. Marconi's vision (and indeed that of many of his rivals, Popov included) was of a system of using radio waves to transmit Morse code over distances, a feat previously only possible with a cable. So in its early days, radio was known as 'wireless telegraphy',[1] which is why some old people will still refer to a domestic radio set as 'the wireless' (something which used to baffle me since the set used by my grandparents in the 1960s was quite clearly connected to its wall socket with a wire. What's more, when I illicitly removed the back of the set with a Meccano screwdriver, it was found to be noticeably full of wires).

The distinction is important, because the concept of broadcasting from one station to countless receivers had not yet emerged. Even so, the advantages of sending dits and dahs without a cable link were obvious, especially where ship-to-shore communication was concerned. By 1898 Marconi had established a factory in Chelmsford, Essex, to manufacture equipment for exactly that purpose, and in that same year Queen Victoria was using Marconi 'wireless' to keep abreast of the health of the Prince of Wales, who was convalescing aboard his yacht. The following year, Marconi was hailed as a life-saver when a ship ran into trouble in the North Sea and the first wireless SOS was transmitted.

His ambition, though, was to send a telegraph signal across the Atlantic using radio waves instead of the established (but often troublesome) undersea cable. Many thought this impossible, believing that radio waves, like light, must travel in straight lines and would thus be lost to space owing to the curvature of the earth.

On 12 December 1901, Marconi proved that in fact radio waves would bounce back to earth from the ionosphere, a hundred miles above the surface. A signal sent by his most powerful transmitter, sited in Poldhu in Cornwall, was successfully received by Marconi himself at St John's in Newfoundland through an aerial suspended in the air under a giant kite.

The first transatlantic telegraph in history was painfully modest – just three dots, representing the letter 'S', and then so weak that it couldn't even be printed on the telegraphist's paper tape. There was much scepticism surrounding the event, so two months later Marconi repeated the experiment with a transmission between Poldhu and a receiver on board the ship *Philadelphia* 1,500 miles away. This time he was able to point to two miles of telegraph tape covered with thousands of messages, the autheniticity of which was certified by the captain and first mate of the *Philadelphia*. 'Will they now say I was mistaken in Newfoundland?' Marconi asked.

An outside broadcast in 1924. Even though he's on the radio, Major J.A. White, reporting, sports his regimental tie and his monocle.

So although Popov is a contender for the title of the Father of Radio, it was Marconi who, by 1904, held the vital patents for radio in the UK and US, and it was the Marconi Company that became the world leader in ship-to-shore communication equipment. Within the first ten years of the twentieth century, wireless telegraphy became a fact of life. But before the end of its second decade, it was time for a fundamental change in the philosophy of radio.

David Sarnoff arrived in the US in 1900, aged nine, when his family finally escaped the poverty of a small Jewish community in Minsk, Russia. He was an ambitious boy, and by the age of thirteen had saved enough money to buy his own news-stand for $200. Later, he secured a job as a messenger boy for Marconi and by absorbing an understanding of wireless telegraphy was able to rise to the rank of junior telegrapher, on $7.50 per week, by the age of sixteen. By April 1912, at twenty-one, he was working in the Marconi station on top of the Wanamaker Hardware Store in New York, processing some of the most infamous ship-to-shore telegraphs of the century.

Marconi's transmitters were already being hailed as the saviour of the 700 people who were rescued from the *Titanic* disaster by ships that had picked up her SOS calls. Now Sarnoff was disseminating news of the survivors for the benefit of the relations who thronged the streets outside.

But Sarnoff wanted to be more than a mere telegraph operator and had ideas of his own about the future of radio. He put it to the men who ran Marconi's empire that while this two-way communication was all very well, what you really needed was a system of transmitting from a single station to many receivers – or 'music boxes', as he imagined them. Like the man at Decca Records who made the mistake of turning down the Beatles, the Vice President of Marconi US, Edward J. Nally, thought there was no future in the idea. Marconi led the world in the ship-to-shore business and was not to be diverted.

But Sarnofff was, so he left Marconi and took his idea elsewhere – to the newly formed Radio Corp of America, or RCA. They gave him $2,500 to develop his scheme. It was on 2 July 1921 that he finally and unequivocally demonstrated the value of radio waves as a one-way means of broadcasting entertainment to the masses. Like all the great media moguls, he recognised that a sporting event was the way to the nation's heart, and his live trans-mission of a prize fight between Jack Dempsey and Georges Carpentier is reckoned to have attracted 300,000 listeners.

But while that might be seen historically as confirmation that radio for the masses was not to be denied, it was by no means the first use of radio waves for speech or music. In 1900 the Canadian Professor Reginald Fessenden sent a voice signal a very shoutable distance of fifteen metres; six years later, on Christmas Eve, he made what is considered the world's first entertainment broadcast from Brant Rock in Massachusetts – a violin performance by Fessenden himself of 'Oh Holy Night' and a short speech. The telegraph operators of ships in the North Atlantic were astonished to hear words and music coming from the equipment that normally yielded

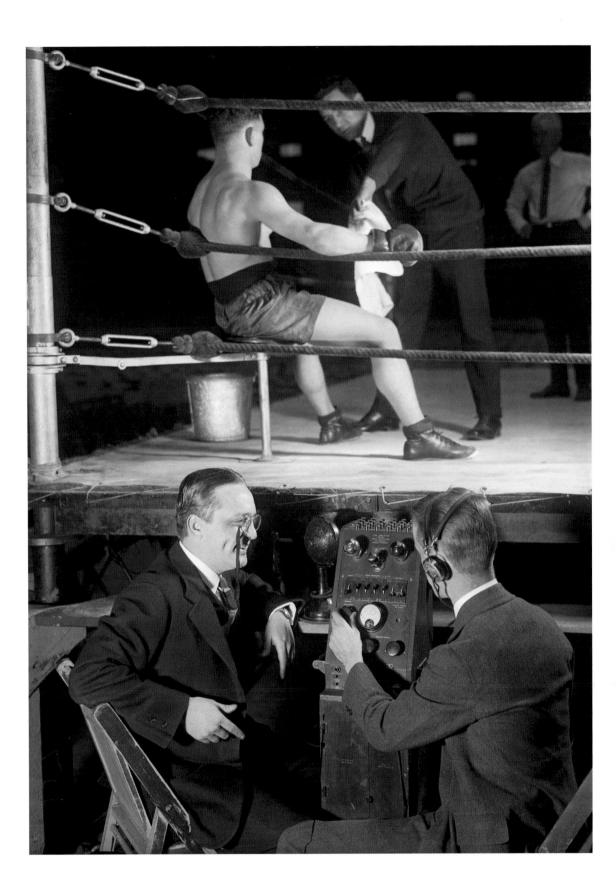

'TO FERMENT
THE TEENAGE
UPRISING, THE
RADIO HAD TO
BECOME THE
SORT OF DEVICE
THAT COULD BE
USED IN PRIVATE'

nothing more than the dots and dashes of Morse code.

The significance of Sarnoff, however, is that he recognised how big radio could be. Prior to him, broadcasting was the preserve of amateurs and hams who perhaps lacked his sense of possible proportion. But at least that meant that when organised broadcasting took off in the second half of the 1920s, there were already plenty of people with receiving equipment.

That equipment was the crystal radio set, the first even vaguely practical means of listening to the radio at home. The crystal radio set seemed to have sprung equally from deep science and witchcraft. It comprised only a handful of basic components – children could, and did, make them – and, most incredibly, needed no power source.

The crystal set gathered radio waves through an unfeasibly long antenna – essentially a thin wire that might be trailed right across a large room. Astonishingly, and provided it was firmly earthed, that same antenna also provided just enough electrical current to drive the device. A tiny crystal, usually galena, was enough to convert the radio signal into something audible, but only through headphones, since the set did not include any form of amplifier.

Making a crystal set work was also something of a diabolical art. The antenna had to be manipulated endlessly to make the most of the signal, and a tiny wire that came to be known as a 'cat's whisker' had to be made to tickle the surface of the crystal in precisely the right way if the radio waves were to be properly deciphered. The receptive acuity of a crystal set was also quite feeble, so though the BBC would one day boast that 'nation shall speak peace unto nation', in the early days of domestic radio it was more a case of tuning into someone maybe ten miles away with an instrument to pluck or a book to read from.

Even so, it is hard to think of another innovation whose social significance is so out of proportion to its simplicity. For the first time in history it was possible to hear people who were out of earshot and who, in all probability, you had never even met!

By the 1920s, radio stations were proliferating. In the first four years of the decade over a thousand were set up in the US, many of them commercial ventures that mixed sponsors' messages with programmes of speech and phonograph recordings. In Britain, a more conservative approach was taken; advertising was considered inappropriate and only a few large private companies were granted licences to transmit. Eventually, a syndicate of broadcasters (made up of companies that manufactured radios, no less) was founded and financed by the Post Office, which levied a ten-shilling fee on anyone who bought a radio. It was this operation that eventually became the British Broadcasting Corporation in 1927.

Despite this, radio caught on in the UK. In 1924 licences were held by 10 per cent of British households. By 1933 this figure had risen to 48 per cent and by the outbreak of the Second World War, 71 per cent.

It helped, of course, if you were reasonably well off, because then you could afford a valve radio, which unlike crystal sets, contained an amplifier, and that meant listeners were freed from the tyranny of headphones and antenna and could listen through a loudspeaker. Valve radios were just more sociable.

But they weren't without their downsides. Valves needed to be hot to work, so warming up the set could take half a minute or more. They also required a hefty power source, and in the days before mains electricity was standardised this meant an expensive battery of 110V or more, the cost of which represented a week's pay for a typical working man.

Valve radios were also heavy; not just because the mass of valves mounted on their steel chassis amounted to quite a bit of hardware, but because they were ornately hewn from stout wood like miniature Art Deco cathedrals. Ostensibly, the valve radio was portable and often equipped with a massive leather carrying strap to confirm as much. In reality, it was a fragile piece of furniture afforded a place of permanence in the sitting room, where it glowed like some New Age hearth as the focal point of the modern family.

So radio was still far from being a subversive force, because it was still firmly rooted in the parental domain of the home. If radio were to be used to ferment the teenage uprising, the radio itself had to become the sort of device that could be used in private, and ideally under the bedclothes. And miniaturising the radio meant miniaturising its single most vital component – the valve.

The valve's replacement was the semiconductor, or transistor; it does the same job, but is as tiny next to a valve as today's silicon chip seems when

placed alongside the soldered circuit board that was its forebear.

The semiconductor arrived in 1947 and was the work of a team of three men working at Bell Laboratories – William Shockley, John Bardeen and Walter Brattain. But history remembers Shockley as its creator, partly because the company he later founded in the apricot groves of California – Shockley Semiconductor – was the first inhabitant of what became silicon valley, and partly because he once employed eight of the world's leading physicists. But mostly because he was prepared to trample on anyone to be remembered as such.

At the press conference where the Bell team's pioneering work was announced, Shockley hogged the microphone and pushed Bardeen and Brattain aside in an attempt to ensure he alone appeared in the historic photographs. Eventually, his despotic style of leadership became so unbearable that his eight physicists left him to set up Fairchild Semiconductor, and in turn Intel, the world's most successful semiconductor company.

The semiconductor was the key to shrinking all kinds of electronic control devices that had once relied on the gently humming valves, and to creating whole new technologies, too. But its significance to radio was so

In the early days, portable radios were expensive to buy. So poor people, like these shoeless urchins, rented them.

James May's 20th Century: Making Waves

enormous that in the new era of the truly portable, hand-held receiver the prefix 'transistor' seemed as definitive as the world 'radio' itself.

A dozen semiconductors could sit on the end of the thumb, whereas a whole hand was needed to hold one or two valves. A transistor radio required much less power than a valve one, and could be run for days on a battery smaller than a matchbox, which in turn made the whole thing even more compact. The new injection-moulded plastic did the job that craft-era woodworking fulfilled on the casing of the domestic set, and suddenly a radio might fit in a pocket.

A transistor radio was seen as an entirely different device, and in many ways it was, because it was used to listen to an entirely different kind of broadcast. That the advent of the personal and portable radio roughly coincided with the rise of rock 'n' roll no doubt helped cement the relationship between the 'tranny' and the first stirrings of teen rebellion, but the liberty to take the radio anywhere you wanted was the key.

Knowing that the kids could be listening away from Mum and Dad meant that broadcasters could produce the type of programme that Mum and Dad wouldn't want them to hear; not just back-to-back pop but frank discussion shows on topics as delicate as sex. Radio remained communal and universal, as Sarnoff had envisaged, but now it had also become exclusive. With the revolution in the design of the radio set came a parallel revolution in the culture of radio itself.

Perhaps it was not as momentous as we have been led to believe. Yes, the image of the recalcitrant teenager with the brightly coloured pocket radio is an alluring one, but in reality it didn't last. The activities of the notorious offshore Radio Caroline were greatly suppressed in 1967 by the British government's Marine Broadcasting Offences Act, although, in fairness, Radio 1, which was launched later in the same year, was modelled largely on Caroline and other 'pirate' stations. But by being a BBC station, Radio 1 confirmed that even kicking against the system had already become mildly institutionalised, at least in its broadcast form.

Radio survives – indeed, has proliferated – as a communal and sociable medium with a permanent location in the home and workplace. And back in 1984, Queen felt its future was looking good:

You had your time, you had the power
You've yet to have your finest hour.

On the other hand, Dinah Washington, in 1953, could see the way things were going:

TV is the thing this year, this year
TV is the thing this year
Radio was great, now it's out of date
TV is the thing this year

Washington was the more prophetic, because by the end of the century

television (in its broadest sense) had been emancipated in the way radio was by the semiconductor revolution. The emergence of the unscheduled and downloadable video clip means that the spirit of the seditious pirate radio station survives on sites such as YouTube. The transistor radio has merely been usurped by the iPod.

Otherwise, nothing much has changed. Fifty years ago, adults were beginning to worry about what their teenage children were listening to. Now it's what they're watching.

Radio girl always got her man.

THE HUNDRED YEAR FORMAT WAR

As well as giving teenagers the transistor radio, the twentieth century also bestowed upon them another gift; the chance to replay their own music in their own homes, their own bedrooms, and their own clubs. Recorded music became a crucial part of a teenager's persona; an opportunity for self-expression that had never existed before.

The record collection became a revealing window on the soul. Imagine a scene from the 1970s: a young man invited back to a young lady's room. While she made coffee or looked in the back of the fridge for a bottle of Blue Nun, it would to customary to fill the awkward silence by thumbing through her LPs. A quick rummage would give you a very good idea of how the rest of evening, perhaps even the rest of your life, might turn out. If her collection was dominated by Hendrix, The Eagles and perhaps a touch of Janis Joplin, the signs were good. If she owned entire cannon of Joni Mitchell, you'd better tread carefully. And if you found yourself thumbing through The New Seekers, Dana and Julie Andrews then it was time to make your excuses and get out.

By contrast modern formats – the DAT, the rewritable CD, the MiniDisc or MP3 – offer relatively little insight about a person's inner self. Collections are no longer carefully built up over years, but can now be ripped from a thousand websites in an evening. How did music, once something so precious and cherished, become a commodity that behaves so promiscuously? Well it all started with a format war a long, long time ago.

Before the 1870s, if you wanted to listen to music in your own home the only viable solution was to learn an instrument or sing. So it's easy to understand why, when the chance to enjoy recorded music came along, it was joyously embraced; even if, by today's standards, it was rubbish.

The first system to be developed was the phonograph, in 1877, by Thomas Edison. It was said to be his proudest invention; quite something when you remember Edison held the world record of 1093 patents, which included everything from the incandescent light bulb to the motion picture camera. The phonograph was both a recording and replaying machine,

and Edison had high hopes for it. He believed it would be ideal not just for recording music, but also for more esoteric purposes, such as preserving threatened languages and recording the last words of the dying.

The main rival to Edison's phonograph was the gramophone, developed by Emile Berliner in 1887. The gramophone recorded music onto flat brittle disc. The disc would spin at 78 rpm and would last about five minutes. No amplifiers existed, so the sound was created mechanically; the grooves on the record would wobble a stylus, which would in turn wobble a membrane to create the sound. An enormous horn, a sort of ear trumpet in reverse, amplified it. There was no real control over the volume other than to stuff something soft into the horn, such as a sock. This is believed to be the origin of the expression 'put a sock in it'.

Edison's rival system worked in much the same way, but its groves were not on a disc, but on a cylinder about the size of a corn on the cob (an essential difference is that on the phonograph the stylus vibrated vertically, while on the gramophone it vibrated laterally). The first words to be recorded with his system were 'Hallo, hallo' followed by a rendition of 'Mary Had a Little Lamb'. The cylinder had a key advantage: the speed of the needle in the grove was constant from beginning to end, unlike on the gramophone disc where the stylus 'moved' quickly to begin with, but as the needle worked its way steadily into the middle on the disc, its speed in the grove would slow down. As a consequence the volume would drop and, worse still, as the groves got tighter and tighter the music started to sound awful.

But convenience won the day; gramophone discs were easier to press and easier to store. Soon 78 rpm discs were everywhere, and despite their rather fragile nature, they can still be found in car boot sales to this day. They barely changed for fifty years, which for a recording format is a very healthy life span indeed. But in 1948 a new tougher and more flexible record came along. It was vinyl, and with it more music could be packed onto the same sized disc. Vinyl polymers enabled 250 micro-groves to be cut into every inch of the disc's radius, far more than the 80 groves per inch of the 78. This, combined with a slower playing speed of only 33 and one third rpm, meant that a staggering 25 minutes per side was now possible – nearly one whole hour on a single disc, which was enough even for a Rachmaninov piano concerto, sadly (In the days of the 78, Rach 2 came in a whole box of short-playing records, an excellent disincentive).

But there was another way of exploiting the excellent capacity of vinyl. One big record could hold a great deal of music; conversely, one big hit would fit on a very small one; viz., the 45rpm single, which was launched in 1949 by RCA Victor. Here was a recorded music format with an ASBO, and one that would become a vital spur to the teenage rampage.

As is so often the case with innovation, timing is everything. Vinyl appeared just as rock 'n' roll was exploding across America and about to make its way to Britain's more conservative shores. Elvis Presley had recorded his early work in the 1950s on 78 rpm, as had Bill Haley and the Comets, who committed the 1954 teenage anthem 'Rock Around the Clock'

No amount of patriotism could save Edison's silly cylinder-based system.

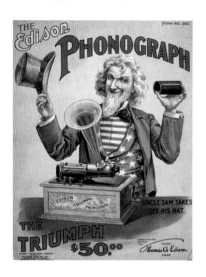

to the mercifully brittle 78. For a short while these artists could be found on
both formats, but that wasn't to last long. Vinyl was to triumph.

Its advantages were pretty obvious from the start. Vinyl was a cheap
material, so singles were pocket-money purchases. They were much easier
to carry around to a friend's house and, compared with 78s, were virtually
indestructable. This led to the development of an electro-mechanical marvel
known as the jukebox, which allowed young people to amass in cafes and
pubs, listen to the hit parade, and plan the overthrow of the establishment.

More importantly, along with the single came a new type of record
player – one that didn't need winding up, just turning up. This was the
Dansette, a beautiful and deeply coveted artefact about the size of an
overnight case and finished in funky two-tone colour schemes like a
contemporary Chevrolet Bel Air. Of course, a single only lasted about
as long as an old 78, but the Dansette had an autochanger arm that fed
a stack of hits onto the turntable in succession, to keep the party going.[2]
A simplification of the jukebox, in fact. It looked as though it had flown
in straight from California, but in fact the Dansette was a combination
of ingenuity from Birmingham and Russia.

Morris Margolin, a Russian, had arrived in the UK in the 1920s and
started a small family business making cabinets. He also had an interest in
musical instruments, and combined his skills to produce a record player – the
'Plus-a-Gram'. It was the forerunner of the Dansette. Then in the early 1950s

the Margolins joined forces with the Birmingham Sound Reproducers (BSR), who made turntables. Their first model was the Dansette Senior. It was to become a massive success, the business became 'Dansette Products Ltd', and soon had a workforce of several hundred, with people queuing up outside their warehouse waiting to buy these cherished record players.

It is hard to imagine how The Beatles or The Rolling Stones could have corrupted the world without the 45 single and the portable record player. But no sooner had vinyl records begun to change the world than boffins were already trying replace them. As early as 1965 there were dark mutterings that the stylus should be dispensed with and replaced by a laser beam, which wouldn't wear the record out. Scratch-free records that never stuck and ruined the moment of seduction would be hi-fi heaven. Indeed, there was a system that scanned conventional vinyl LPs with a laser, but it didn't catch on. Lasers would, though.

Meanwhile, in 1963, Phillips came up with a more immediate challenger to the vinyl record, the Compact Audio Cassette (or Musicassette, or simply Cassette Tape, or just Cassette). It was intended as a means of improving the sound quality of dictation machines, but its appeal to cash-strapped youths was obvious. For a start it was, like Edison's Phonograph, a recordable medium and would eventually be available in durations of two hours, which meant three cassettes could provide the soundtrack for an all-night party, given some judicious borrowing of other people's records. The music business was as worried about this as it would later be about Napster.

Cassettes stretched, attracted dirt, jammed, became tangled in the mechanisms of their players and sometimes unraveled entirely to form miles of brown bunting, but to their credit the machines that played them could sustain very rough treatment. This meant a cassette player could be fitted to a car in a way a record turntable never could (unless you were John

Yet more record-player-induced gay abandon.

The Sony Walkman offered a life of pure hedonism interrupted only by mangled tape.

Lennon, who had a turntable, floating on a bed of oil, fitted to his Rolls-Royce Silver Cloud), and without the cassette tape the personal stereo would have been delayed by another decade.

But while bootleggers were busy arguing over the merits of chrome tape and Dolby Noise Reduction, Philips were already perfecting something much better. Work on the CD was begun in 1975 by Lou Ottens. Sony was in partnership with Philips and legend has it that Akio Morita, Sony's boss, stipulated that a CD must be able to store the whole of Beethoven's 9th Symphony. Thus CDs were developed to play for 74 minutes.[3]

So a CD was longer than an LP record, which meant that when old LP albums were re-released on the new format, there was room for more; for bonus tracks, out-takes and other things that, in the age of vinyl, had been rightly consigned to the dustbin of pop. But since CDs cost more, it was essential to convince buyers of their 'added value'.

Of real value was the resilience of CDs. The recording was no longer a true recording at all, but an encoding of sound in a digital stream of noughts and ones, read by a laser beam and recreated in the electronics of the player. CDs made even vinyl singles look delicate. When *Tomorrow's World* demonstrated the CD in 1981, the presenter tried to ruin the disc by smearing something like jam on it, and still it played perfectly – no clicks, no background noise, no fluff on the needle. (It now seems that *Tomorrow's World* smeared the top side of the disc, not the playing side. But nobody knew that at the time.)

And, as with cassette machines, CD players could withstand knocks without malfunctioning, which made a portable, personal CD player a possibility. It wouldn't have happened, though, without the example of the cassette, which was the 'enabling technology' behind the first Sony Walkman. Once again Akio Morita was the driving force, a man who liked to listen to his Beethoven in private. At first the idea was dogged by corporate uncertainty – would anyone want a tape machine that didn't record and had no speakers? And what should it be called? Stowaway, Freestyle and Soundabout were all suggested, but it was 'Walkman' that found its way into the Oxford English Dictionary in 1986, having attained the status of a generic term a mere seven years after the launch of the original device in the summer of 1979 (one key to the success of the Walkman was its small headphones. They would only get smaller). Fifty million were sold in the first 10 years.

The technological leap from a cassette Walkman to a CD Walkman was therefore a relatively straightforward one, and with the establishment of digital recording the century could close to the faint chirrup of a commuter's MP3 player. Finally, every stumbling block of music recording had been overcome: portability, affordability, durability, sound quality, recordability. Even the storage problem had been overcome.

To frame it in the sort of double-decker buses, football pitches and distance-to-the-Moon type of analogy that has become so popular, the music on an iPod would, if transferred to 78 rpm records, fill a shelf 100 feet long.

INSTRUMENTS OF THE DEVIL

'FENDER'S GUITAR LOOKED LIKE IT HAD BEEN FASHIONED FROM THE WINGS OF A CORVETTE STINGRAY'

Jimi Hendrix and his white Strat at the Royal Albert Hall. Nobody did it better.

Teenagers, and indeed the rest of us, may have been acoustically liberated in the twentieth century thanks to transistor radios, personal stereos, CD players and MP3s, but a recording medium on its own was never going to transform the world. At the same time as music recording was changing, so was music itself. Once again, new technology lay at the heart of the revolution.

Prior to the twentieth century, musical instrument-makers had not been wholehearted in their embrace of new technology. The violin is regarded by many as having reached its apotheosis in the work of Antonio Stradivari (1644–1737), and the tone quality of a Strad is generally attributed to the density of the varnish used. The flute underwent a radical rethink during the early nineteenth century, in the hands of Theobald Boehm, who devised a new key system for it and substituted metal for wood in the body of the thing. In the 1840s, the Belgian Adolphe Sax gave us the saxophone, but in truth his instrument was something of a failure. It was intended to have a serious and high-brow position within the orchestra, but eventually got a job on the Chattanooga Choo Choo.

So at the start of the twentieth century, instruments were pretty much as they had been for a very long time. Consider the guitar, which had not changed significantly since its invention (the guitar is an ancient instrument which, contrary to popular belief, predates the lute). If you wanted to play one, the acoustic version was the only model in town.

The acoustic guitar was revealed to be pretty hopeless in the era of jazz, because it was too easily drowned out by the massed brass of a big band. One solution was to place a microphone directly in front of the instrument, but while this amplified the guitar, it also amplified everything else in the vicinity, including other instruments and the player himself if he collided with his mike during an enthusiastic solo.

A better solution was the electric pick-up. In a pick-up, coils of very fine wire are wrapped around magnets positioned immediately below the guitar's strings. When a string is plucked, the changing flux of the magnetic field induces a tiny current in the coil proportional to the velocity of the string's vibration, that is, its pitch. This signal can be amplified exclusively, free of corruption from pesky trumpets and other usurpers.

Early electric guitars, however, were merely conventional hollow-body guitars to which pick-ups had been affixed, so called electro-acoustics. It followed, though, that since the body of the guitar was no longer responsible for amplifying its sound, as in a violin, it needn't be hollow at all. The guitar could now be solid and virtually inaudible until plugged in, and it was a short step from there to one shaped like a starburst (and, in turn, from there to a swimming pool shaped like a guitar, presumably).

Creation of the solid-body electric guitar is credited to Adolph Rickenbacker who, in the early 1930s, built a cast-aluminium instrument dubbed the Frying Pan. The Russians had something similar, the Kuznetsov electromagnetic guitar, which was displayed at the Exposition of National Economy Achievements in Moscow in 1935. But the Stalin regime, who were generally against people having fun, put a stop to it. Elsewhere, a gifted guitarist and electronics whizz who sometimes went by the stage name of Rhubarb Red came up with something called the Log. This was, in

The young Les Paul, aka 'Rhubarb Red', ready to change the world of rock.

Guitar Wars: Keith Richards wields his Fender Telecaster (top), while Jimmy Page brandishes a Gibson Les Paul.

essence, just a wooden post with a guitar neck attached to it, though at first glance it looks like a conventional guitar because Rhubarb added two detachable hollow-body halves, perhaps to avoid offending the sensibilities of more conservative strummers.

But interesting though the Log and the Frying Pan were, what was really needed was a range of guitars that were reliable, affordable and attractive. The need was particularly urgent as America was under threat.

The threat came from the ukulele. The 1940s saw the popularity of this irritating toy instrument spread like wildfire across the continent. By the early fifties the ukulele menace was epidemic, with an estimated 3 million of the four-stringed novelty items in use across the country, and American citizens at constant risk from busking. It was an invasion, it needed to be stopped, and Leo Fender, a man who couldn't even tune a guitar, let alone play one, led the fight back. He created the simple, bolt-together Fender Telecaster in 1950.[4] It was the first solid-body guitar to be made in significant numbers. This was a real, no-nonsense instrument that was to be favoured by country and western players, and quite a few blue-collar legends: Bruce Springsteen, Keith Richards and Status Quo all play guitars from the Telecaster family.

The early fifties became the golden age of guitar manufacture, and Fender's rival, Gibson, quickly countered with their own solid guitar, inspired once again by the innovative Rhubarb Red. Fortunately for guitar history, and Gibson's marketing effort, the new instrument was quickly christened with Rhubarb's real name, which was Les Paul.

It was a beautifully crafted instrument with two pick-ups, one near the guitar bridge to give a sharp bright sound, another close to the neck to give a gentle mellow tone.

Not to be beaten, Fender responded with what was to become the most recognised electric guitar in the world; it boasted three pick-ups, but more crucial was the way it looked. In contrast to the Gibson Les Paul, which was relatively traditional in shape and appeared to be related to a mahogany sideboard, Fender's guitar looked like it had been fashioned from the wings of a Corvette Stingray and became available in bright automobile colours: Surf Green, Dakota Red or Lake Placid Blue Metallic. It was named the Stratocaster.

So just four short years, 1950–4, saw the creation of the three most important guitars ever built, with the Strat and the Les Paul fighting neck and neck for domination. But this wasn't a format war with one winner and the loser consigned to history; both these instruments have remained dominant to this day, and so has the argument over which is best. Guitar heroes have divided themselves up quite fiercely: Hendrix a Strat man, Jimmy Page more Les Paul; Eric Clapton, Strat; Peter Green, Les Paul; Tony Blair, Strat. Nearly every serious guitarist will have owned one or other in their career, sometimes even giving them a name; Clapton called his Strat Blackie, George Harrison's was Rocky. Blackie was recently sold at auction for $959,500.

The electric guitar was one of the most important breakthroughs of the twentieth century, a weapon of social reform as much as a musical

Parfitt and Rossi of Status Quo explain how to play the second of their three chords.

instrument. Perhaps Stalin could see its power when he tried to stifle it at birth. So could Jimi Hendrix, who said: 'I wish they'd had electric guitars in the cotton fields . . . a whole lot of things would have been straightened out.' Never mind that for most teenagers who picked up an 'axe', volume, feedback and the effects of wah-wah pedals and fuzzboxes were a ready substitute for the real talent that would be needed to master the piano. It was something their parents hadn't had and therefore didn't like, so it was better than the piano, which was a piece of furniture in Grandma's house.

Rock 'n' roll would not have got far without electric guitars, but for rock to become truly progressive it needed something far more esoteric, and a bit more ridiculous: an instrument capable of creating sounds never heard before, created by a serious boffin.

One such man was Lev Sergeivitch Termen. He was the boffin's boffin, a man who made his mark in both electronic music and espionage, a man whose life reads like a pulp fiction novel. Born in St Petersburg in 1896,

Termen was an electrical engineer, a cellist and an astronomer. His most famous invention was an electrical instrument that actually used the human body as part of its circuitry. Originally the by-product of Russian government-sponsored research, the first model was invented in 1919 and was called the thereminvox, but it soon became known as the theremin (Termen later became known in the US as Theremin).

The theremin is unique amongst musical instruments in that to play it you don't actually touch it. You simply stand behind it and wave your hands at a couple of antennae, looking for all the world more like a mystic or magician than a serious musical performer. The body's ability to store a tiny electrical charge (its capacitance) alters the electromagnetic field around each of the antennae and in doing so controls the pitch of the note and its volume.

Termen played his theremin for the Russian Bolshevik leader, Lenin, at the Kremlin in 1922. Remarkably, he wasn't despatched to the gulag. It is reported that Lenin was so impressed with the device he considered taking lessons himself, and commissioned 600 of the instruments to be sent throughout the Soviet Union. Termen was despatched abroad as a techno ambassador, demonstrating the genius of Soviet technology to the world.

After a tour of Europe, where he played the instrument to packed halls, Termen set sail for the US to perform with the New York Philharmonic. He then settled in the city and set up a laboratory where he continued to develop the theremin and finally patented the instrument, before granting commercial rights to RCA, who sold about 500 instruments in the 1930s. But the theremin, like the saxophone, failed to make the grade as a classical instrument; instead it was seen as a novelty act, and could be found playing the music halls. It was claimed that anyone who could hum or sing could play the theremin, but this turned out to be arrant nonsense. Most people can produce nothing more than corny woo-woo-woo ghostly sound effects with it.

The theremin was taken more seriously by the composers of film scores. A theremin can be heard in Hitchcock's *Spellbound*, and in sci-fi classics like *The Day the Earth Stood Still*. Some recordings were made, in particular 'Music out of the Moon' in 1947. The record sleeve warns that 'music can affect the sensitive mind in a way that is sometimes frightening'. Neil Armstrong, someone not easily frightened, took 'Music out of the Moon' on his journey to the real thing in 1969.

Termen stayed in America for less than a decade before returning to the Soviet Union under rather suspicious circumstances. At the time no one was quite sure why he left, but it later became known he had been taken back to the USSR by Soviet agents. There he was imprisoned on the orders of Stalin, along with other scientists who had annoyed the Soviet despot, and spent his days developing listening devices for spies. Without Termen around as ambassador, the appeal of the theremin quickly waned and by the 1950s it looked as though the days of bizarre electronic instruments were numbered. (The theremin does live on, and can be heard in the music of Led Zeppelin, Badly Drawn Boy, Goldfrapp and Portishead.)

Just in time, a new saviour appeared, Robert Moog. Moog was a native of New York City and clearly a bright chap. He attended the Bronx

The well-dressed Leon Termin (aka Theremin) and his remarkable instrument.

Robert Moog attached a piano keyboard
to a telephone exchange to create the most
influential instrument of the 20th century.

High School of Science, graduating in 1952, after which he earned
a bachelor's degree in physics from Queen's College and a degree
in electrical engineering from Columbia University. His subsequent
PhD was in engineering physics from Cornell University.

Moog was building theremins while still at high school, and by the
age of nineteen had founded a company building theremin kits to be
assembled at home. Clearly, here was a man destined to become the
darling of Tangerine Dream and the likes.

In 1964 he created the Moog synthesiser which, unlike the theremin,
featured a conventional keyboard. That, however, was the only conventional
thing about it. The Moog combined banks of oscillators, wave generators
and filters to create completely new sounds that doubtless satisfied the
contemporary desire for things that seemed to have come from outer space. It
could produce notes so deep it could perform a duet with a humpback whale.

As a practical instrument, it was the stuff of nightmares, however. It
was the size of a wardrobe and looked like a telephone exchange, with
wires hanging from it like errant spaghetti. It was not, in fact, intended as a
performance instrument at all, since it took hours to set up and was virtually
impossible to manage. Instead, it was regarded as a sophisticated studio

Walter (aka Wendy) Carlos and his (her) synthesiser.

instrument. There it was played with great effect by Walter Carlos (now Wendy Carlos – he eventually synthesised himself) who made two albums in the late sixties: *Switched-On Bach* and the *Well-Tempered Synthesizer*. The first earned Carlos three Grammy Awards. But the sounds the Moog created, and its space-age looks, meant it was not to remain trapped in the studio for long and it was quickly to become one of the most striking instruments to adorn any rock stage.

The Moog opened the floodgates for many other synthesisers, but it was not until the introduction of digital technology in the late 1970s that synthesisers shook off their esoteric image and moved into the mainstream. The breakthrough came with the Fairlight CMI, the first commercially available digital sampling instrument, much loved by Peter Gabriel and Stevie Wonder. The Fairlight could do far more than create new sounds. It could readily capture existing sounds and manipulate them in ways never before heard, whether it was an aboriginal conch shell or a breaking window. The modern synthesiser was the only serious rival the electric guitar has ever faced in pop music. Now you can buy samples of any instrument you want, straight off the shelf, including, ironically, some sounds of the early Moog synthesisers.

From the seventies onwards electric guitars and digital machines, now generally lumped together as 'keyboards', monopolised pop and rock. Where once a whole orchestra or big band was needed to fill a concert hall or club, now a handful of disenfranchised youths could deafen a 100,000-seat stadium with just a handful of instruments.

It has been said that the devil has all the best tunes, which is not quite fair, as God has all of J.S. Bach. But as many late twentieth-century parents will testify, Beelzebub has the highest volume.

DOING THE TWIST

'IN FACT, MARLON BRANDO WEARS A HAT THAT FREDDIE MERCURY WOULD HAVE REGARDED AS TOO CAMP, AND LEANS ON A CAFE BAR DRINKING SODA'

Few characters in the history of cinema have cut a less threatening dash than Marlon Brando in *The Wild One*. Here was a supposed delinquent, the leader of a dangerous motorcycle gang, a new youth institution that, like rock 'n' roll, threatened to topple the very pillars of established society.

In fact, Marlon Brando wears a hat that Freddie Mercury would have regarded as too camp, and leans on a cafe bar drinking soda. There have been bicycle touring clubs more menacing than Johnny Stabler and his boys, and OAP coach parties that have done more damage to the roadside services.

Still, *The Wild One* has a special place in the collective hearts of bikers, because it stands as ancient evidence of motorcycling as a cult rather than a mere means of transport. Other bike-inspired movements would emerge: the Hell's Angels, the rockers, the cafe racers, the mods on their scooters. In the 1970s, British teenagers would be driven wild with lust by the crackle of a 50 cc two-stroke moped. Some pubs and roadside cafes would – still do – owe their existence to the ferocious loyalty of motorcyclists and motorcycle clubs. A motorcycle admits its owner to a gloriously inclusive worldwide fraternity in a way the car never can.

But it wasn't always so. The motorcycle, like the bicycle that inspired it, was popularised largely by people who couldn't afford cars.

Like the car, the motorcycle was a nineteenth-century invention that came good in the twentieth. Sylvester Roper of Massachusetts built the first one in 1868, based around the boneshakers of his day and powered by steam. It was not for the faint-hearted. Directly under its seat was a small, vertical, charcoal-fired boiler, which powered a two-cylinder engine to drive the back wheel. Remarkably, it featured a twist-grip throttle, a means of control that has never been bettered. His motorcycles were demonstrated to delighted crowds at fairs all over New England.

Roper, like so many motorcycle pioneers to follow, died in the saddle, or soon after leaving it. In June 1896 he turned up at a track near Harvard with his latest machine. After clocking a speed of 40 mph he came off, and

An early Harley-Davidson. They haven't changed much since.

was found to be dead at the scene. The autopsy revealed that he had not actually been killed in the accident but had suffered a heart attack. He was seventy-three.

In Europe, too, and at almost exactly the same time as Roper's efforts, steam power had been allied to the bicycle, notably in the French Michaux-Perreaux Velocipede. But low power and operational difficulties meant that steam bikes would never catch on in the way steam cars and lorries did.

So it was left to Gottlieb Daimler and Wilhelm Maybach to build the first proper motorcycle in 1885. Their creation was considerably less elegant than the earlier steam bikes, and was horribly upright. It was named the 'Reitwagen', or riding car, which has led some to pronounce it the first car rather than the first motorbike. It did, after all, have four wheels – the two wheels we'd expect on a bike, plus two small outriding stabilisers. More importantly, it used an internal combustion piston engine, as all bikes would henceforth, with the exception of a few rotary-engined types.

The new century was just a year old when the first globally dominant national motorcycle industry, that of Britain, began to evolve. In 1901 Norton, Royal Enfield and Ariel made their first motorcycles, and at the London trade show of that year, a hundred different models were displayed. The following year, Triumph – which had been making bicycles since 1889 – also produced a motorcycle. Essentially, it was a Triumph bicycle with a small engine attached to its frame's downtube, but 500 were sold within a year. This, despite the suggestion in a popular technical magazine that the novelty would soon wear off.

Meanwhile, in America, the three Davidson brothers – Arthur, Walter and William – teamed up with William Harley to build motorcycles. Although the marque they founded is now considered the definitive American bike, they were at the time overshadowed by their rival Indian, whose trademark deep red bikes would be the world's best-seller by the outbreak of the First World War, with 32,000 sold. At the time, there were around 600 bike manufacturers in the world, a number that would dwindle very rapidly.

The Second World War did no great favours to the progress of the motorcycle, just as it didn't spur the development of the car. The military certainly wanted bikes and lots of them, but they wanted simple, rugged, soldier-proof designs that could be cheaply made and easily maintained. The First World War put many small bike-makers out of business.

A better incentive to develop bikes came from racing. The Isle of Man, the most famous and hazardous bike circuit in the world, held the first of what would become the TT series in 1907. (Car racing came even earlier to what has been called 'the road racing capital of the world', with the 1904 Gordon Bennett trials.) The TT would become a barometer of nations' dominance in motorcycle building. Britain ruled on the island until the years immediately after the Second World War, when the Italians moved ahead. They, in turn, were vanquished by the new bike giant from the east, Japan. As in the TT, so in real biking life, it would seem.

Real biking life, however, was a lot less glamorous. Although the motorcycle has come to be seen as something of a leisure vehicle, most of its riders throughout history have braved its hazards, its poor weather protection and its downright impracticality simply because they couldn't afford anything better. The motorcycle was the natural progression from the bicycle that had taken the world by storm in the late nineteenth century. A bicycle was cheaper than a horse; a motorcycle was cheaper than a car. In Britain, between the wars, motorcycle numbers matched or exceeded those of cars. In developing countries today the natural progression from bicycle to motorcycle to car is easily observed – in India and China, for example – and this explains why some of the most significant bikes in the history of motorcycling have been, as with the 'people's cars', fairly modest affairs. But more of that anon.

Indian was once more famous than Harley-Davidson, but has now been largely forgotten.

Overleaf:
British ton-up boys en route to a fight and a series of accidents, some time in 1964.

'BIKE GANGS
FOUGHT WITH
EACH OTHER, AND
THE "ROCKERS"
ON THEIR BIG-
BORE BRITISH
BIKES WOULD
EVENTUALLY
GO TO WAR WITH
THE "MODS" ON
THEIR ITALIAN
SCOOTERS'

The working man on his bike faced a problem if he married and produced children – how to transport the family. One solution was the sidecar. The idea had been around since 1903, and they were used extensively in the Second World War (Germany was especially good at sidecars. On some the 'third wheel' of the sidecar was driven along with the rear wheel of the bike proper, which was useful in rough or muddy terrain). A 'combination' (as a motorcycle and sidecar is correctly known) could transport up to four people, at least if two of them were still quite small. The irony is that the fitment of a sidecar denies a motorcycle the attribute for which it is most admired today, namely its ability to cut through traffic.

Since many people were restricted by circumstance to riding motorcycles, it was inevitable that biking would turn into a movement. After the Second World War, returning American servicemen looking to rekindle the camaraderie of the battlefield formed motorcycle groups and gave birth to the spectre of *The Wild One*. The idea of the 'biker', a restless man who embraced motorcycling as a way of life rather than just as a means of commuting to work, was created.

The small machine with the BIG PERFORMANCE

Even in the 1950s, sex was being used to sell blazers and tennis racquets.

Left:
The mods – Italian style, British attitude. It could have been worse. It could have been the other way around.

Something similar was happening in Europe, and especially in Britain, whose motorcycle industry was the envy of the world. Great classics from Norton, Triumph and BSA were the mounts of the 'ton-up' brigade, those riders who could achieve the magical 100 mph at a time when most cars struggled to crack 70 mph. The sport of 'cafe racing', in which riders set times for the journeys between their favourite roadside haunts, accelerated the art of engine tuning and motorcycle modification almost as much as the racetrack did. Motorcycling was rebellious, heroic, and had its own dress code of leather, insignia and tattoos.

But there was a dark side. The motorcycle and motorcyclists' attire became associated with trouble. Bike gangs fought with each other, and the 'rockers' on their big-bore British bikes would eventually go to war with the 'mods' on their Italian scooters, terrorising the seasides of Britain and filling the front pages of newspapers with tales of Viking-like destruction and pillage.

The casualty rate among motorcyclists was horrific, too. For decades, the improvements in frame, suspension, tyre and brake design lagged behind those achieved with the engine, with inevitable results. To the modern rider, motorcycles from the fifties, sixties and even seventies are terrifying not because of how fast they go – not very, by today's standards – but because of how slowly they stop.

For the fledgling teenage biker, eager for freedom, two-wheeled life would always start with something simple. Britain gave its youth the BSA Bantam, which stayed in production from 1948 to 1971, its engine growing from 123 cc to a spine-tingling 173 cc. It enjoyed a reputation for being dependable, enduring, and easy and cheap to maintain, the motorcycling equivalent of something like the Austin 7 of the unattainable car world. But the best small bikes have been produced by countries that were forced by fortune into relying on them, and the two greatest small bikes in history are Italy's Piaggio Vespa scooter and Japan's Honda Cub.

Both were conceived as simple, universal transport for the people of countries ravaged by war and short on materials and manufacturing capacity.[5] First seen in 1946, it was the response of a young aeronautical designer called Corradino d'Ascanio to a request from the boss for 'a machine to put Italy on wheels'. D'Ascanio didn't actually like motorcycles much, so brought to the drawing board the thinking of the aeroplane designer. In profile, the resulting Vespa was more than redolent of the wheel spat of an aeroplane's fixed undercarriage and the front suspension was really a simplified oleo leg. The engine was hidden in the bodywork and drove the rear wheel directly – d'Ascanio especially distrusted chain drive. Most importantly, the Vespa was designed to be ridden by anyone, even people in skirts, which in Italy included priests. The Vespa, though not strictly the first step through scooter, became the blueprint for every scooter since and has been licence-built around the world.[6]

Honda's Cub, the work of Soichiro Honda himself, arrived twelve years later but was even more portentous. Though inspired to some extent by the Vespa and other step-through scooters, it was even more ruthlessly simple. The frame, as on many early Honda bikes, was of pressed steel,

Top:
The Yamaha FS1-E, the greatest motorcycle ever produced.

Bottom:
The Honda Cub, the motorcycle produced in the greatest number.

Right:
The Honda CB750, death knell of British bikes.

to which simple plastic leg shields were added. The engine was a feeble but bullet-proof four-stroke 50 cc single. The Cub, in its various guises, became the most numerously produced motor vehicle in history, and while exact production numbers are difficult to ascertain, the figure is certainly approaching 40 million.

The Cub was a dire warning to the dominant but complacent British motorcycle industry, the decline of which has become an object lesson in bad management for all other manufacturing businesses. The British bike-makers, their companies already greatly consolidated, acknowledged that Japan would become the leader in small bikes but believed they could simply concentrate on the high-value and prestigious big-bike market. Japan thought otherwise. The bike that most resonantly sounded the death knell of British bikes was another Honda, the CB750 of 1969.

The big Honda had everything British bikes couldn't offer: four cylinders, a disc brake at the front, and indicators as standard. Most remarkably, its horizontally split crankcase did not leak oil, while virtually every vertically divided British engine did. The Honda could be said to have kick-started the age of the Japanese superbike, were it not that it was fitted with another unimaginable convenience, an electric starter. Within ten years the names of Honda, Yamaha, Suzuki and Kawasaki would usurp Triumph, Norton and BSA on the wish-list of dreaming British schoolboys.

In Britain, the Japanese makers even allowed teenagers to cement their loyalty to the new eastern deities at an especially tender age and, inevitably, they did it with a small bike. A largely forgotten piece of legislation said that while the minimum age for a motorcycle was seventeen, a sixteen-year-old could ride a Motor-Assisted Pedal Cycle, or Moped, with an engine of 50 cc or less. In the post-war years, 'moped' generally meant exactly that – a bicycle to which a rudimentary clip-on engine had been added, often to drive the front wheel. The Japanese revived the concept but with one important difference – they built proper 50 cc motorcycles and simply fitted pedals to comply with the law (other countries' bike-makers saw the sense in this, too, especially in Italy, but never with the success the Japanese enjoyed). The result was the 'sports moped', the most famous of which was Yamaha's FS1-E, or 'Fizzie'.

Great claims were made for the Fizzie and its ilk, though more often by their owners than their makers. Since sixteen-year-olds could still be at school, playgrounds were stunned by mutterings of 60, 70 and even 80 mph. In reality, sports mopeds struggled to crack 45 mph, but no matter. To sixteen-year-olds, to whom the legal age for a real motorcycle was an eternity away, the moped was an electrifying improvement on the bicycle. More importantly for the manufacturers, the thousands of teenagers who began motorcycling on a Japanese moped would probably stay on a Japanese bike for life.

Today, surviving European makers such as the revitalised Triumph, Ducati, Aprillia and BMW are small-volume players. Harley-Davidson continues as an occult religion in its own right. The vast majority of motorcyclists are riding Japanese.

But in the developed world, the heyday of the motorcycle is in many ways over. Rising standards of living and lower prices have seen car sales soar and motorcycle sales gradually decline. Even the young have largely abandoned two wheels, making the leap from childhood bicycle to car in one painless step. Apart from scooters, the small and medium-sized workaday motorcycle is in the doldrums, and most machines are exotic and expensive recreational toys. What's more, the high price of these bikes, the prohibitive cost of insurance for young riders, and the ever more complex requirements of the motorcycle licence have seen biking turn into something of a middle-aged pursuit.

The fraternity is as strong as it ever was, perhaps stronger. But its devotees are even less threatening than Brando.

THE NEW FABRICS OF SOCIETY

In the 1950s nylon promised a glamorous future where everyone would look this good.

Snobbery is a terrible thing, and takes many terrible forms. Social snobbery is as old as society itself, and has developed into a mechanism for controlling it. There are music snobs, literature snobs, food snobs and even type-of-computer-hardware snobs. But one type of snob –perhaps the most perverse type – could not exist until the twentieth century: the materials snob.

In an age of organic produce and faux craft traditions, when the pretentious and aspirational are desperate to attain respectability by championing small-volume 'exclusivity' and a flaws-are-part-of-its-charm approach, it is tempting to be aloof about Formica, PVC, polythene, nylon, polystyrene and the many other synthetic fibres that have been conjured out of a test tube. But how would life be without them? These materials deserve to be cherished because they are, quite literally, supernatural.

Back at the beginning of the twentieth century, there was an astonishing lack of colour and textural variety in the world. Unless you commanded a high position in society, and had access to silk, gold and rubies, fine art and exotic rugs from the east, your life would look as drab as it undoubtedly was. If you were poor, waiting for spring and picking a flower to stick in your hair or the lapel of your only suit was probably the best opportunity you had to brighten up your otherwise turgid appearance. Back then, the materials that made the world were steel and iron, wood and stone. Clothes would mostly be the product of plant life, stolen from a passing sheep, or fashioned from the by-products of butchery. There was precious little scope for self-expression; certainly no easy way to create the multicoloured, vibrant, exciting world we enjoy today. We know from our grandparents that in their time people had very few clothes, so Carnaby Street could not have existed, except as a row of shops selling identical grey suits and brown smocks.

But before new materials with magical properties could colonise Planet Earth some serious chemistry needed to be done, and some molecules with seriously long formulae needed to be created. The beginnings of this process can be traced back to the late 1860s, when a competition in America quite literally set the ball rolling: it offered the massive prize of $10,000 to anyone who could find a substitute for ivory, specifically for billiard balls.

The material that looked most promising was called Parkesine. It had been developed by a Brummie chemist, Alexander Parks, in 1862, and is regarded as the first genuine plastic. (Vulcanite predated it, but that wasn't a plastic. It was a means of modifying rubber.) Parkesine was romantically described as a 'beautiful substance for the Arts', mainly because it was displayed at the London International Exhibition of 1862 as a series of decorative pieces. It would be more normally used for making humdrum odds and ends, such as combs and other knick-knacks for the home.

An American, John Hyatt, thought Parkesine might win him the billiard-ball prize, so he altered the formula and created a new material, celluloid. Apart from the small matter of its volatility, celluloid looked up to the job. Composite billiard balls were coated with Hyatt's new formula and were put into play. Thus a new dimension was added to the genteel game of billiards – when two of Hyatt's balls collided there would be a sharp explosive sound, not unlike that of a pistol being fired. This was not conducive to a relaxing game, particularly in the more trigger-happy states of America.

Another even more unlikely use for Parkesine was to make false teeth, which proved cheaper than the existing rubber dentures. Unfortunately, they tended to soften when drinking very hot tea, and they gave the tea a strange taste.

The first person to make a genuinely synthetic plastic was Leo Baekeland, a Belgian cobbler's son no doubt driven to despair by the unrelenting leatheriness of the world he moved in. Baekeland had a way with chemistry, and – unusually for an inventor – a way with money. He had already made it rich in New York by inventing a photographic paper and selling the idea for $1 million. Still only in his thirties, he could have retired happily, but instead kept on working and in 1907 developed the first synthetic polymer. It was the forerunner of modern plastics. Its decisive quality was that it could be moulded when heated, but when cold would set permanently hard. It was resistant to both heat and water, and so made a great electrical insulator. This made it ideal for radios and televisions, as well as cameras and clocks. It became known as 'the material with a thousand uses' and was called Bakelite; so Baekeland, who was already a very rich man, became a very, very rich man, eponymously celebrated in the description of old radio knobs for evermore.

Baekeland allowed companies in Britain to develop his idea, for a fee of course. One was called the Damard Lacquer Company of Birmingham – it is believed they chose the name because the material was 'damn hard'.

All this began a rush for synthetic materials: neoprene was created in 1932, polythene in 1933, perspex in 1934. But while this might be altering the way your kitchen looked, or how aircraft were built, so far synthetic materials had made precious little impact on clothing, except, perhaps, to effect a shift of modern fashion from society parties to the riverbank, where neoprene waders became all the rage. The man to change all that was Wallace H. Carothers, described as a 'brilliant artist in the manipulation of invisible worlds'.

In the 1930s, if you wanted to wear fine stockings, you had to rely on a little chap called the mulberry worm. This creature, which looks a bit like a small white caterpillar, led a simple enough life, ate a lot, and generally minded its own business – the production of silk thread, which can be woven into silk cloth. For centuries, silkworms had toiled tirelessly in the Far East to provide luscious silks for emperors, aristocrats and, of course, lovers of silk stockings.

But in America, scientists at DuPont spotted a business opportunity and were determined to put the silkworm out of business. This promised a rich reward; 1930s America was very heavily dependent on the Japanese silk industry, as each day a staggering 1.55 million silk stockings were sold.

It was no small commitment by DuPont. It was to cost $27 million and take eleven years of research to defeat the silkworm, but eventually their new wonder material was created. Here it is, and you were warned:

$$n\,H_2N(CH_2)_xNH_2 + n\,Cl(C=O)(CH_2)_y(C=O)Cl \longrightarrow$$
$$[\ldots(NH(CH_2)_xNH(C=O)(CH_2)_y(C=O))_n\ldots] + 2n\,HCl$$

Real silk stockings – soon to be a thing of the past.

James May's 20th Century: The New Fabrics of Society

1945, and a Mrs Ruth Cregan can't wait to get home and try on her new nylons.

Left:
How to sell nylons, California 1950: the actress Marie Wilson is hoisted skywards to admire the two-ton cast of her own leg.

This is the formula for the humbly named hexadecamethylene dicarboxylics. Legend has it that when the first fibre was pulled from a glass beaker in the DuPont laboratory, the scientists were so excited that they ran around the lab, covering it with a giant web of the new material, just to see how far it would stretch.

This would be the future of snappy dressing, but not while it had such an un-snappy moniker. If the formula had been hard, choosing a name would be even harder. Many were suggested, and rejected.

It is believed that DuPont originally intended to name its fabric Norun; that is, a stocking that wouldn't run. But someone must have been far-sighted enough to realise that one day several billion women would be available to testify against them in court, and it was turned down. Another potential name, at least according to lore, was Duparooh, which meant, to those who knew, DuPont Pulls A Rabbit Out Of a Hat. Fortunately, they chose nylon.

Some argue that 'nylon' is a contraction of New York and London, where the chemists working on the project came from, or that the name was taken at random from a New York to London air ticket, where the route was abbreviated to NY-LON. It is of little consequence. Nylon was such an incredible breakthrough that DuPont could have called it almost anything, except by its proper name.

It was an innovation of epic proportions. Now people could wear something that hadn't come out of a worm's bottom or grown out of the ground. And because nylon would initially be used to make cheap stockings, it was incredibly sexy.

DuPont spent a small fortune in promoting nylon as an exciting 'modern marvel'. It was first shown to an eager public at the World's Fair in America in 1939, worn on the shapely legs of Miss Chemistry, who emerged from a giant test tube. Nylon finally went on sale on 15 May 1940, or N-Day, as it was known. DuPont sold a staggering 750,000 pairs within hours. That year, they produced 64 million pairs of stockings.

So nylon was ready to take on the world. Unfortunately, so was the terminally unfashionable Adolf Hitler. This meant DuPont stopped its nylon stocking production to respond to the other more pressing needs of the Second World War. Nylon was used for, among other things, parachutes, ropes, tents, flying suits, tyres and – not widely known, this one – blood plasma filters. As nylons became more of a rarity, a black market began to operate; stockings, normally a dollar a pair, began to fetch twenty times the price, and as American servicemen stationed in Britain often had a pair or two stashed away, they became renowned as a means of seduction. Visiting American airmen with winning smiles and a pair of nylons helped fuel the baby boom in East Anglia.

In Britain, stockings were only available to foreigners. Shoppers were required to establish their foreign credentials with a passport before they were allowed to buy them. Nylons were so valuable that during the investigation into a particularly nasty murder in Chicago, one possible motive, robbery, was ruled out because the killer had left six pairs of nylon

stockings at the scene of the crime. No villain with money on their mind
would have done such a thing.

In August 1945, a mere eight days after Japan's surrender, DuPont
announced to a grateful world that it would immediately return to producing
nylon stockings. This was the beginning of the nylon riots. DuPont could not
make the stockings fast enough. It was reported that fights broke out when
women, who had been queuing for hours to buy nylons, were turned away
empty-handed.

It took six months for DuPont to accelerate production back up to the
point where it would balance demand. By then the range of nylon clothing
had been expanded to include nylon dresses and nylon lingerie. Cotton
became as square for clothing as wood was for the casing of radios.
Nylon, largely reviled these days as a low-rent substitute, was the stuff
of progress and modernity. Synthetics were replacing natural materials
everywhere – in household appliances, in the interiors of cars, in furniture,
in carpets, and even in personal effects such as jewellery and briefcases.
Clothing would be no different.

It wasn't just that nylon could be produced very quickly and was
hence relatively cheap. It could also be coloured like no material before it,
and optimistic colour was craved in the post-war years. Nylon was also
easy to maintain, easy to wash at low temperatures, resistant to staining
and easy to iron, if you bothered at all. Drip-dry represented liberty. The
triumph of nylon in the world of teenage fashion was not just because
it allowed a rapid wardrobe turnover at relatively low expense, it was
because, like all teenage clothing, it was invisibly marked with a 'Mum
wash only' label. Nylon was as low maintenance as its wearers were high.

Wallace Carothers was never to enjoy watching the revolution he had
started. He killed himself with cyanide in a fit of depression in a hotel room
in Philadelphia in 1937, before his invention came onto the market. He
never saw the astonishing outcome of his work; that a bit of white goo that
came out of a pot was to become inextricably linked to sexiness and youth
and fashion.

Hugely encouraged by the power of nylon, DuPont, and many others in
the chemical industry, began to look for more and more synthetic materials
that they believed would be appeal to a new, novelty-hungry and – most
importantly – richer type of young person. ICI's Crimplene was hailed as
the fibre success story of the sixties. Man-made fabrics actually began to
influence the shape of modern style – the definitive stay-pressed slacks of
the seventies would not have been possible without Polyester.

And without Lycra, youths of the eighties and nineties would not have
been able to step out in what looked a bit like their PE kit. Lycra had also
been invented by DuPont, back in 1959. It was initially known as Fiber K,
and was intended to banish elasticated seams to the rag bin of clothing
history. Stretchiness is built into Lycra's very soul: it has the comic-book
capability to expand to five times its original size and then snap back to
shape uncorrupted (technically, Lycra is not a fabric at all but a synthetic
elastic). It is one of the most promiscuous of materials and will blend with

almost anything – silk, wool, cotton, even leather, without affecting the look and feel of its host fabric.

So where did it all go wrong? At some point, the headlong rush to embrace synthetics resulted in their use in applications where they just weren't suited. An obvious example is nylon bedlinen, which endured long beyond the point where incessant nocturnal electrocution revealed it to be truly nasty, and that occurred almost immediately. There is also no doubt that 100 per cent synthetic underwear could restore BO to the heights it had achieved before the new automatic washing machines were invented. The backlash was bound to be extreme.

But in truth, synthetic materials never really went away. They are still abundant in mass-market consumer fashion; in bags, straps, sportswear, the trimmings and linings of coats and, most evidently, in waterproofs. However, changing attitudes forced nylon, polyester, rayon and crimplene to collude with the natural fabrics they were supposed to usurp, and adopt a more clandestine role. Now, an ostensibly cotton or woollen garment will often include a smattering of man-made material, there to modify the shape, add colour or simply increase durability.

Synthetics, like everyone and everything else, know their place.

…but even after a further 20 years, they could do nothing for the men.

WAR AND THE WORLD
PLANES
TANKS
ESPIONAGE

5

THE FIGHT FOR
THE HIGH GROUND

The much-reviled Royal Aircraft Factory BE2.

Left:
A Nieuport 17 dives on a flight of German
Albatross scouts some time in 1916.

Standing next to a surviving Royal Aircraft Factory BE2 of 1915 (there's one in the Royal Air Force Museum at Hendon, north London), it is easy to see why the early aviators arrived at the word 'kite' as a term of endearment for their machines. The BE2, with its taut fabric, wooden framework and wire bracing, seems to owe a great deal to things normally seen flying on the end of a piece of string.

There are even more terrifying examples. Four years before the BE2, the Bristol Aeroplane Company, Britain's first true aeroplane factory, had built a particularly cumbersome-looking machine and named it the Boxkite. To have called it anything else would have been frankly deceitful.

Despite all this, the BE2, along with many of its contemporaries, exhibited certain flying qualities that would perhaps delight many a jaded present-day flyer. Low power – just 70 hp in early versions, courtesy of a Renault engine – necessitated a large wing area and a small wing loading (that is, the weight supported by the wings expressed as a load per unit of area, usually pounds per square foot in Britain). It was very light at 621 kg, or 65kg less than the original Austin Mini. So the BE2 offered excellent 'buoyancy', would take off at a gentle trot and was very stable. It as much floated as flew around the sky, and the influence of the wind was as big a consideration as the pull of the puny engine, making it the sort of craft in which the operator must enter a pact with nature and physics instead of trying to overcome them. The BE2 invites the accolade, 'That's proper flying.'

But it is easy to be romantic. In truth, the early aeroplanes were treacherous mounts. The science of aerodynamics and airframe design was in its infancy, and development was inevitably priced in crashes. The vital fabric covering of the wings – the substance of the aeroplane's aerodynamic attributes, since without it the wings were mere frames – could peel off in high-speed dives, and the wings themselves could fold up in violent manoeuvres. The stall, spin and loop were demons, and many never recovered from their confrontation with them. And the engines were horribly unreliable.

'THE BE2
EXHIBITED
CERTAIN FLYING
QUALITIES
THAT WOULD
PERHAPS
DELIGHT MANY
A JADED
PRESENT-DAY
FLYER'

Engine failure was not such an issue at altitude, because aeroplanes like the BE2 could glide superbly. But it was more likely during the full-power stress of take-off, when poorly trained pilots would attempt to turn back to the airfield, stall, spin and crash.[1] Even today, light aircraft pilots are rigorously taught that, in the event of a power failure on take-off, a landing must be made *straight ahead*.

If the engine ran, it was still a menace, at least if it was one of the rotary types such as the French Gnome. These engines employed a total-loss lubrication system in which oil was thrown from the cylinders as a mist after the combustion stroke. As the oil was castor oil – and therefore a laxative – and was breathed in by the pilot, many flyers fought a continuous battle with their bowels long before they encountered an enemy aircraft. The solution was believed to lie in the vigorous consumption of beer.

Finally, once aerial combat began, a further horror emerged: the ever-present risk of fire. The engines and fuel tanks of early aeroplanes were usually mounted close together, in the interests of mass centralisation and to simplify the fuel-feed system. There was none of the armour plating or self-sealing that would be in place for the next air war, only a red-hot exhaust pipe waiting to touch off a tinderbox of well-seasoned wood and fuel-soaked fabric – the ideal materials for a good bonfire. The early pilots, as is well recorded, had no parachutes. It is true that in Britain some people in authority believed that parachutes were an affront to moral fibre, and that pilots equipped with them might feel inclined to abandon aeroplanes that might otherwise be saved. More importantly, the early parachutes were incredibly bulky, and simply didn't fit in the snug cockpit. (Observers in balloons generally did have parachutes. They had the luxury of space and they were defenceless.)

That defines the type of machine in which the first air war of history was campaigned: fragile, dangerous and largely misunderstood, both by those who flew them and those charged with their deployment.

Remarkably, the BE2 briefly held the British altitude record at just over 10,000 feet, a height at which, any modern flyer will tell you, the risk of hypoxia is considerable. The courage of the aircrew is beyond compare. They were venturing into a new and uncharted arena of combat, in a new type of weapon, writing a new codex of warfare as they went. But the reasons for being up there were as old as warfare itself.

Two and a half thousand years ago, the value of elevation in warfare was already well understood. In his classic text, *The Art of War*, the Chinese strategist Sun-Tzu advised warriors to 'observe on high ground and face the sunny side'. (In 1940 this maxim was revised into a caveat by RAF Fighter Command: 'Beware of the Hun in the sun.') For the next two millennia, Sun-Tzu's high ground was a hill or a tower.

The aeroplane would offer very high ground, and movable high ground at that. Balloons had already demonstrated how useful this was; the aeroplanes added predictability. The first tentative use of an aeroplane for military reconnaissance came in 1912, and with some success, when

An early aerial photograph showing the First World War trench system between Loos and Hulluch.

Captain Piazza of the Italian army took a Bleriot XI aloft to see what the enemy, the Turks, were up to in Tripoli, and even photographed them.

Even so, before the outbreak of the First World War, there was still much resistance to the aeroplane from the military establishment. They couldn't see a practical use for it, despite painstaking demonstrations by the pioneers during, for example, manoeuvres on Salisbury Plain. But by 1914 Europe's military leaders had, rather grudgingly, accepted that this new and infernal flying machine might just be able to take on part of the work generally discharged by the cavalry: reconnaissance.

This mindset – that the aeroplane was simply an observation platform – had a profound effect on the development of early military machines. It was thought, reasonably enough, that the business of recording the enemy's movements would task the crew so greatly that they should not be over-burdened with the job of controlling the aeroplane. A reconnaissance aeroplane should be as stable as possible – should fly 'hands off', as pilots would put it.

Apart from anything else, they would soon be required to take aerial photographs. The cameras of the time were cumbersome affairs; large, heavy and, like everything else in the olden days, made of wood. 'Taking a picture' involved the fiddly exposure of delicate glass plates, which was especially tricky in the bulky clothing and thick gloves needed to withstand the low temperatures at high altitude. As a general rule, air temperature decreases by 2° C with every 1,000 feet increase in altitude.

(On a freezing winter's day the temperature, even at 5,000 feet, could be lethally low.)

So benign flying characteristics were the order of the day. Britain produced its BE2, Germany the Rumpler Taube, France the Maurice Farman MF-11. Aeroplanes bumbled around the sky in their passive role as aerial voyeurs, their crews untroubled except by the vicissitudes of the machines themselves. But soon enough, the vulnerability of such aeroplanes to enemy 'scouts' made them extremely unpopular with their pilots. In Britain's Royal Flying Corps, especially, the BE2 and its successor, the equally stable RE8, were pig-headedly kept in front-line service long after their failings had been exposed. Arthur Gould Lee, an RFC pilot writing in 1917, said, 'Everyone is glad to see death-traps like Quirks (Quirks was the RFC flyers' poignant nickname for the BE2) written off, especially new ones' (in the First World War more British airmen died in the BE2 than in any other type).

At the beginning of the war, the crews of reconnaissance machines could do little more than wave at one another, and often did. They would even photograph each other with the equipment that was supposed to be pointed at the trenches. But this wouldn't last long. Aerial reconnaissance was already too important to be allowed to go on unchallenged – by the end of the war Germany would be processing 4,000 photographs a day.

Aerial photography wasn't easy as the cameras – like the aircraft – were still very primitive, as this US Navy model shows.

Above left:
An FE2d offers one solution to the problem of firing through the propeller. It is a pusher design with the freedom to shoot in all directions.

Above right:
Roland Garros modified his Morane-Saulnier Type N by fitting metal deflector plates to the propeller blades – just visible in this picture – to enable him to fire forwards.

Something had to be done, so bricks were thrown, and pot-shots were taken with rifles and revolvers, with very little result.

A machine-gun, that newly proved and most efficient weapon of war, was obviously what was needed on an aeroplane, but even this idea was met with some resistance by officialdom. The annals of early air warfare are peppered with stories of innovators being carpeted by senior officers for wasting time trying to arm aeroplanes. When the idea was finally accepted, it still presented problems. Early gunners would often occupy the forward cockpit of a two-place machine such as the BE2, for reasons concerned with preserving the centre of gravity. From there, they were perfectly positioned to shoot up the wings of their own aeroplanes.

Later, more powerful machines such as the legendary Bristol Fighter would place the gunner in the back, with a good field of fire. In the early days, however, one ingenious solution was effectively to reverse the fuselage, putting a pusher engine at the back between a twin-boomed tail and placing the gunner in the nose. The Vickers FB5 Gunbus was so arranged.

It would be better, though, if the gun could be rigidly fixed to the airframe and the pilot could aim the whole machine as part of the act of flying. Then the aeroplane would become a weapon in its own right, rather than simply a glorified tripod for the mounting of one. The problem now was the propeller, which was in the way.

The British solution would be another pusher type, the single-seat DH2 with a machine-gun fixed on the nose. France would build the Nieuport Bébé, on which the gun was mounted on the top wing, above the propeller's arc. But the German answer came first and was the most influential: the

The Fokker Eindecker, armed with its
synchronised machine-gun, was rightly
feared by Royal Flying Corps pilots.

The Fokker Eindecker, armed with its
synchronised machine-gun, was rightly
feared by Royal Flying Corps pilots.

Fokker Eindecker, a single-seat 'tractor' monoplane with, remarkably,
a machine-gun that could fire *through* the propeller.[2]

Both British and French engineers had experimented with a synchronised gun but, as usual, their efforts had been dismissed by blinkered
authority. In desperation, Roland Garros, a French pilot (the first man to fly
across the Mediterranean), fitted deflector plates to the propeller blades of
his Morane-Saulnier Type N in order to deflect any rounds that might
otherwise destroy it, although at some personal risk from ricochets. So
armed, he shot down three German aeroplanes before he was brought
down behind enemy lines by ground fire.

The Germans showed Garros's machine to Anthony Fokker, a brilliant
Dutch aeroplane designer in their employ. Within three days – at least
according to legend, though he had probably been working on it already –
he had designed his synchronising mechanism, which allowed a Spandau
machine-gun to fire safely through the propeller of his Eindecker. The
outcome was devastating to the Allies, many of whose pilots would now
end their brief flying careers in a moment of pure astonishment as an
enemy aeroplane fired at them from the one position where it shouldn't be
possible. The German aces Max Immelmann and Oswald Boelcke flew the
E1, and formed part of what quickly became known in Allied circles as 'the
Fokker scourge'. BE2s and the like were 'Fokker fodder'.[3]

The Fokker gave Germany immediate and devastating air superiority;
the victims were largely the unwieldy and too-stable reconnaissance
aeroplanes. Britain's DH2 and France's little Nieuport would eventually be
a match for the otherwise rather primitive Eindecker. The Allies regained
the initiative with, from Britain, the Sopwith's 11/2 Strutter, Pup and
Triplane, and from France the enlarged and improved Nieuport 17.
Germany's response was the beautiful (and often beautifully decorated)
Albatross series of biplanes, which in turn spurred the development of
Britain's Sopwith Camel and Royal Aircraft Factory SE5, and France's
Spad series, later beloved of American flyers in the Escadrille Lafayette.

The air was now a true field of combat, and a new type of aeroplane
had arrived to contest it – what we would now call the fighter. By the end of
the war there were some magnificent single-seat fighting machines in
service – Britain's late-coming Sopwith Snipe, Germany's Fokker Dr1
triplane and Fokker DVII, to name a few. The fighter pilot had been born,

Above:
SE5a fighters of 203 Squadron in 1918: strong, powerful, agile, well-armed, possibly the best fighter of the First World War.

Below right:
The Fokker Dr1 Triplane, immortalised by Manfred von Richtofen, the Red Baron.

too; a man who prized agility, speed, fire-power and strength in an aeroplane, not the lumbering, stolid characteristics of the BE2.

Dogfighting, which had been inconceivable at the start of the war, was now the principal spur to the development of aeroplanes and their engines and armament. A comparison between the BE2 of 1914 and the Sopwith Camel of 1917 is illustrative: the BE2 had a top speed of around 72 mph, took at least twenty minutes to climb to 10,000 feet, and carried no armament. It was inherently stable. The Camel could fly at 118 mph, took half the time to reach 10,000 feet, and carried two synchronised Vickers machine-guns. It was manoeuvrable to the extent that it could catch out the unwary.

The aerial duellists who flew the new fighters have been portrayed as dashing knights of the air with a strong sense of chivalry. There is doubtless something in this, but it is also doubtlessly true that the air war in 1914–1918 became a gruelling, bloody and unforgiving theatre. Baron Manfred von Richtofen, the Red Baron, once said, 'When I have shot down an Englishman my hunting passion is satisfied for a quarter of an hour.' He collected grisly souvenirs from the wreckage of his victims' aeroplanes and awarded himself trophies for his kills. The British ace Mick Mannock, on hearing of the baron's death on 21 April 1918, said, 'I hope he roasted the whole way down.'[4]

So by the end of the First World War, two possible roles for the hitherto widely dismissed aeroplane were firmly established – reconnaissance and aerial combat. But there was a third and obvious duty that could be performed by aeroplanes in war, since they spent so much time above the enemy's positions: dropping bombs on him.

The first recorded experiment in aerial bombardment actually took place in the world's first true aeroplane, when two American army officers lobbed a couple of small bombs from a Wright Flyer. Early attempts at bombing in the First World War were similar, and very ineffective. Apart

Top:
By 1915 bombers were regarded as a serious threat, as this public warning poster shows.

Bottom:
A pair of Handley Page heavy bombers near Dunkirk in 1918.

Right:
A German observer demonstrates very low level bombing techniques. The aeroplane is actually on the ground.

from the simple fact that the armed bombs were a greater threat to the aeroplane carrying them than to anyone on the ground, early aeroplanes were far too feeble to carry a decisive ordnance payload.

An alternative to explosives was something called the flechette, essentially an oversized steel dart about four inches long. Handfuls of these were tossed out of the cockpit when over the enemy's trenches, and they could reach a velocity sufficient to penetrate a steel helmet (unless perhaps a British flechette met the spike of a German helmet on a directly linear trajectory, in which case a sort of spike stalemate would result). Some members of the British High Command thought flechettes deeply unethical.

Soon enough, means were devised by which small bombs could be hung on proper racks below the wings and released by simple cable systems. Observation aircraft and the sturdier fighters could be armed with a couple of small bombs and used to attack enemy airfields and the like. Italy and Russia led the field in dedicated bomber designs, the Russians producing the vast Ilya Mourometz with a payload of 1,150 pounds, and Italy the Caproni Ca-3, which could carry 1,000 pounds. Both of these machines appeared in 1915, and testify to how seriously the two nations were pursuing this new form of terror.

In Britain, Handley Page built the O/100, a monster of 100-foot wingspan powered by two Rolls-Royce Eagle engines of 250 hp each. But the bomb load was disappointing, so it was improved with uprated Eagle VIII engines of 360 hp each, which increased its bomb load to some 1,600 pounds. The O/400, as it was now known, was used with some success against German military targets in occupied territories; more portentously, at the end of the war it was deployed against the industry of the Rhineland.

Germany, however, was to become the master of the true heavy bomber in the last two years of the war. Its Zeppelin airships had demonstrated the psychological effect that a bombing raid on London could have, even if they caused little real damage. Zeppelin also built the enormous R-VI, which joined the Friedrichshafen G-III and Gotha G-VIII in attacks on Paris and London. The R-VI in particular was an awe-inspiring aeroplane with a wingspan of over 138 feet, or just three feet shorter than that of Boeing's B-29 Superfortress of the next war. In typically literal German style, the R-VI was known as the Riesenflugzeug, or 'gigantic aeroplane'. The smaller Gotha (wingspan 78 feet) was known as the Grossflugzeug – 'large aeroplane'. The German air force included a distinct bombing wing, the Bombengeschwader.

So tactical bombing, by small aeroplanes carrying small bomb loads that could be directed at specific targets, was an accepted part of the art of aerial war by the close of hostilities in 1918. The concept of strategic bombing, in which much bigger aeroplanes carrying much bigger bomb loads were launched against more general targets, often at night, arrived too late to change the course of the conflict to any degree. But the concept seemed a sound one; it would be revisited and eventually someone, as Britain's 'Bomber' Harris predicted, would reap the whirlwind. ('They have

The Supermarine S6B, winner of the 1929 Schneider Trophy Race and a spur to the development of future fighters, notably the Spitfire.

sown the wind, they shall reap the whirlwind.' He was actually quoting the Old Testament – Hosea 8:7.)

At the outset of the Second World War, warplanes were already an entirely different proposition from the types that had concluded the conflict just two decades earlier. In this war, air power would be confirmed as a force in its own right; as a sole means, some believed, of winning.

So what had peacetime done to uphold the giddy pace of aeroplane development? Initially, not much. The victorious nations were spending their money on more constructive things, and the defeated were essentially banned by treaty from producing military aircraft.

But on the quiet, things were going on. Germany soon began to encourage aerial activity that would translate readily enough into aircrew training for war – gliding, and sport flying in aeroplanes such as the Bucker Jungman. The new airliners and fast mail planes under development required only the addition of guns and bomb doors for a change of role from world-shrinkers to destroyers.

In Europe and America, the hugely popular spectacle of air racing provided a ready proving ground for engines and airframes. The greatest air race on earth was the Schneider Trophy contest for seaplanes, which began in 1913 and was drawing crowds of around 200,000 by the late twenties. Any nation that could win the race three times within five years would become permanent custodian of the hugely ornate trophy itself. It was the perfect arena in which to develop a new breed of fighter.

Curtis of America and Macchi of Italy were strong contenders: both would go on to build superb fighters. Britain fielded machines from a Southampton-based flying-boat builder called Supermarine, which won the trophy outright in 1931 with its Rolls-Royce-powered S6B.

The S6B was an outstanding aeroplane that, shortly after its Schneider victory, set a new world speed record of 407.5 mph. More importantly, it allowed its designer, Reginald J. Mitchell, to formulate much of the thinking that would go into a future project, and the racing experience helped Rolls-Royce to develop the engine that would power it. Mitchell's triumph was the Supermarine Spitfire, powered by the Rolls-Royce Merlin.[5]

The Spitfire entered RAF service in 1938, a year after the dowdier but more rugged (and initially much more numerous) Hawker Hurricane. Around the world, other nations were pursuing similar thinking in fighter design, and at roughly the same time. America had its Curtiss P-40, France its Dewotine D-520 and Morane-Saulnier M.S.406, Italy the Macchi Folgore, the USSR the Yak 1 and Germany the outstanding Messerschmitt, Bf 109 (the most numerously produced fighter of the Second World War, over 35,000 being built), later described by one RAF pilot as 'ever present, always dangerous'. Japan, a new air power, also had a great fighter, the Mitsubishi Zero.

All were sleek, powerful monoplanes with enclosed cockpits. The build-up to the inevitable war finally ended the era of the front-line biplane fighter, and with it went some of the most charismatic aeroplanes ever built: the Gloster Gladiator, the Bristol Bulldog, the Fiat C.R. 32 and the

'MITCHELL DID NOT CHOOSE THE NAME SPITFIRE HIMSELF; THE WAR MINISTRY DID. HE IS REPORTED TO HAVE SAID THAT IT WAS "A BLOODY SILLY NAME"'

James May's 20th Century: The Fight for the High Ground

Grumman F3F, to name just a few. The biplane layout would endure in trainers, some amphibians and, most famously, in British carrier-borne torpedo bombers such as the Fairey Swordfish (the Stringbag, as it was known, was responsible for crippling the *Bismarck*, so it can't have been all bad). But the fighter could no longer be seen as a gentleman's duelling weapon. It was a lethal instrument of total war.

And since the end of the Second World War was also the beginning of the jet age, the piston-engined single-seat fighter is generally seen as having attained its apotheosis by 1945. Contenders for the title of best piston-engined fighter of all time might include the Hawker Sea Fury, the Focke-Wulf FW190 Ta152, the Vought F4U Corsair and the Grumman Bearcat.

Fighter tactics would change, too. The lightness and extreme agility that characterised biplane types would be eschewed in favour of speed and rate of climb, since altitude as well as outright power could be converted into vital speed for a quick strike and a quick escape. Armament increased, from typically two machine-guns to four or eight, combined with cannons firing explosive shells. Dogfighting was no longer the main objective, because the fighters had a more important threat than the enemy's fighters to deal with. 'Forget the fighters,' yells the Spitfire squadron leader in the 1969 epic *Battle of Britain*. 'It's the bloody bombers we're after.'

The bomber's development was never going to be as glamorous as the fighter's, even though there was plenty of bombing theory expounded during the inter-war years. The British had successfully used small bombers to subdue rebel tribes in Iraq. Germany had more recent experience in the Spanish Civil War, where its Condor Legion had flattened Guernica and introduced the world to the spectre of 'terror bombing'. It was believed that 'the bomber would always get through', and that bombing alone could bring a nation to its knees.

Precision dive-bombing as a means of battlefield support soon proved devastatingly effective, especially when delivered by Germany's feared Junkers 87 Stukas.[6] Diving almost vertically, with the air-brakes extended, the Stuka pilot could simply aim his whole machine, rather like the First World War pilot with his synchronised machine-gun, and be confident of hitting quite a small target such as a tank or gun position. The slow and lightly armed Stukas would ultimately prove easy meat for Allied fighters, but more sophisticated aeroplanes such as the Hawker Typhoon would be striking at vital targets with bombs and wing-mounted rockets up to the war's end.

The aircraft carrier, an idea in place in the First World War but worked out thoroughly during the inter-war years, meant that small precision bombers could be launched close to the enemy. The Americans excelled at a new breed of tough, radial-engined, carrier-borne dive-bomber and torpedo-bomber that would prove vital in the island-hopping Pacific war against Japan. Japan, too, understood the value of torpedo-bombers and dive-bombers, and demonstrated them at Pearl Harbor.

But the heavy bomber and its associated carpet bombing had yet to materialise. In 1939 virtually all bombers were twin-engined, poorly defended and ill-equipped for navigation at night or in fog. They were designed to be used in daylight, against specific tactical targets, and under the escort of fighters.

The Battle of Britain tested the theory, with Germany carrying out daylight raids against airfields and factories, using the twin-engined bombers it had deployed so successfully in Spain. It helped that the British

Left:
The terrifying Junkers Stuka dive bomber.

Right:
The formidable Hawker Typhoon could carry rockets or bombs under its wings.

Opposite:
The gun camera in a de Havilland Mosquito records the final moments of a Junkers Ju 88 over the Bay of Biscay.

Overleaf:
A Heinkel He III bomber over Rotherhithe in London's East End, 7 September 1940.

defenders had Radio Direction Finding – or radar, as it would later be known – which had been developed by the Scottish physicist Robert Watson-Watt after he had been asked by the Air Ministry to investigate the viability of a radio-based anti-aircraft 'death ray'. The death ray was a ludicrous idea, Watson-Watt concluded, because radio waves would never be powerful enough to destroy an aeroplane. But they could detect one. Early detection meant that the Spitfires and Hurricanes of Fighter Command could be scrambled in time for them to gain the altitude that was becoming so vital for a successful interception.

The defending fighters showed that the bomber wouldn't necessarily get through, and the meagre armament of Germany's Heinkel III, Dornier 17 and Junkers 88 proved unequal to the task of fending off a determined attack by an eight-gun fighter with a margin of speed in excess of 100 mph. (Nevertheless, the Junkers 88 remained an excellent and versatile machine, serving as a bomber, dive-bomber, torpedo-bomber, reconnaissance aeroplane and night fighter.)

The British experience was no more encouraging. Early daylight raids against targets in occupied Europe met with heavy losses and rarely inflicted significant damage. The RAF's later twin-engined types, such as the Bristol Beaufighter and, in particular, the de Havilland Mosquito, proved the worth of a twin-engined light bomber that could operate in daylight. The Mosquito was one of the greatest fighting aeroplanes of the war, and worked as a night fighter, bomber, ground-attack aircraft, and spy plane. It could be heavily armed or completely unarmed, relying on its excellent performance to outrun enemy interceptors. But the lumbering British bombers of the early war – the Vickers Wellington, the Handley-Page Hampden, the Bristol Blenheim – were no better able to account for themselves than Germany's Heinkels and Dorniers. The bomber, after all, seemed better suited to night-time operation.

But at night another problem was revealed; or, rather, obscured – that of hitting the target. Truly accurate bombing, even in daylight in perfect flying conditions, was difficult enough. To bomb successfully in the dark, the target would have to be very big. The RAF's Bomber Command proved as much with its early night raids. Arriving within even twenty miles of a factory or railway marshalling yard was a great achievement.

But a whole city could be located at night, as the Luftwaffe proved when it switched its offensive to London and began the Blitz. The East End of the capital was packed with warehouses and factories, but also with houses. The bombing was indiscriminate. Britain retaliated with raids on Berlin, and the destruction of cities became an accepted objective of the war.

Britain became much better at it, partly because it built more aircraft better suited to the task. The four-engined Stirling, Halifax and Lancaster could carry huge bomb loads and could be organised into relentless streams of 1,000 or more. With the help of flare-dropping pathfinder aircraft and early radio navigation aids, they could destroy a city in a few nights, as they proved at Hamburg and Dresden. Germany responded with radar-guided night fighters that inflicted gruesome casualties amongst the

Top:
The Messerschmitt Me 262, the first operational jet fighter of the Second World War.

Bottom:
P-51D Mustangs of 375th Fighter Squadron. Note the underwing long-range fuel tanks, essential for bomber escort duties.

Opposite:
500lb bombs released from a US Flying Fortress fall towards an Italian oil refinery.

unseeing bomber crews. (Heinz-Wolfgang Schnaufer, Germany's top-scoring night fighter ace, downed 121 bombers.)

The Americans, stationed in Britain, had a different approach. They believed in precision bombing, in daylight, from high altitude and in aeroplanes heavily armed against fighter attack flying in tight defensive formations. Its Boeing B-17 Flying Fortress literally bristled with defensive armament: heavy-calibre machine-guns protruded from turrets and perspex blisters for the length of its fuselage. American bombers were equipped with the highly complex Norden bombsight – essentially an electro-mechanical computer – which was so secret it was removed from the bombers overnight and stored in a locked vault. Crew were instructed in how to destroy the Norden in the event of a forced landing in enemy territory.

But the American bombers suffered for the very reasons the German ones had in the Battle of Britain. The enemy fighters were still faster and better armed, and near the end of the war Germany would deploy its Messerschmitt 262 jet fighter and Messerschmitt 163 rocket interceptor against American formations. The accuracy of the Norden sight has been questioned – collateral damage is nothing new – and the weight of a B-17's defensive guns and ammunition gave it a typical bomb load only a third as big as an Avro Lancaster's. American losses were often unsustainable on early raids deep into Germany, and only the development of long-range escort fighters – such as North American's P-51D Mustang and P-47 Thunderbolt – enabled the daylight campaign to continue.

And it was, in the end, an American bomber that delivered the ultimate in indiscriminate bomb loads on 6 August 1945, destroying a whole city at a stroke in the process. The Boeing B-29 Superfortress was the most advanced bomber of the war: pressurised, extremely fast and capable of bombing from 30,000 feet. When the B-29 Enola Gay flew to Hiroshima to drop the first atomic weapon to be used in anger, it did so completely unopposed. Maybe the bomber would always get through after all. And maybe only one was needed.

The jet engine had been under development since the early thirties, by Frank Whittle in Britain and Hans Von Ohain in Germany. It entered front-line service, in the Me 262 and Gloster Meteor, too late and too few in number to change the course of the Second World War,[7] but it was ready to fire a transformation in aeroplane performance during the Cold War that followed.

The first barrier that the jet engine had to overcome was initial top brass indifference, especially in Britain. The second was a performance obstacle, the so-called 'sound barrier': the speed of sound. The term 'sound barrier' has fallen out of use, but was apt at the time since many believed it was precisely that. Aerodynamic phenomena at trans-sonic speeds meant that it could not be breached. It was described as a 'brick wall' and by airmen as 'a big farm you can buy in the sky'. ('Buying the farm' was their euphemism for being killed in a crash, as in 'Ginger's bought it.')

Britain believed the sound barrier could be overcome, and during the war commissioned the experimental and extremely futuristic Miles M52.

James May's 20th Century: The Fight for the High Ground

Top:
The Lockheed SR-71 spy plane.

Bottom:
The innovative Hawker Harrier
operating from HMS *Hermes* during
the Falklands conflict in 1982.

The project was cancelled, but not before much vital data relating to its design, in particular that of its tailplane, had been given to the Americans as part of a research-swapping deal (the Americans reneged on the arrangement, causing much controversy). This data found its way to Bell Aircraft, who by 1946 had come up with the strangely familiar-looking X-1. The X-1 was a rocket-powered aeroplane, but the Miles would have had a proper jet engine.

So the X-1 was the first aircraft to break the sound barrier in level flight, achieving Mach 1.06 on 14 October 1947 in the hands of Captain Chuck Yeager. Later experiments with a one-third scale model suggested that the Miles would have achieved Mach 1.5.

With the imagined aerodynamic impediment demolished, fighter performance was set to soar. The hugely efficient jet engine overleapt another limiting factor of the piston-engined fighter, namely the unassailable dynamic limitations of its propeller and the ever-increasing bulk and complexity of its engine. By the mid-sixties the Lockheed SR-71 Blackbird spy plane was capable of Mach 3 at 80,000 feet, from which position it could survey the earth's surface at the rate of 72 square kilometres per second.

But the fighter remained a fighter, a high-speed interceptor armed with machine-guns, cannons and, increasingly, air-to-air missiles. It was not until the 1970s that the concept of a true multiple-role combat aircraft was developed, notably in the Panavia Tornado (known during its development as the MRCA – Multi-Role Combat Aircraft). Classics of the fifties and sixties include the MiG-15, the Hawker Hunter, the North American F86 Sabre, the Dassault Mirage, the McDonnell Douglas F4 Phantom, the English Electric Lightning and the radical Hawker Harrier.

The advances were not denied to the bombers. Enola Gay and Bockscar (which dropped the Nagasaki bomb) had shown that one aeroplane could do the job of hundreds if armed with a nuclear weapon. So the bomber fleets of the nuclear powers shrank, though not by as much as might be imagined, and the bomber itself grew in stature. One factor was that the early nuclear weapons were very bulky. Another was that they might have to be delivered to very distant targets.

America and the Soviet Union, the main protagonists in the nuclear stand-off, built huge bombers with very long ranges to transport their arsenals. America had the Consolidated Vultee B-36 Peacemaker and the Boeing B-47 Stratojet, and with them established the Strategic Air Command, the nerve centre of its defence against the Soviet nuclear threat. Its headquarters in Nebraska, built in 1948, was then out of range of a Soviet strike. By 1953 it boasted over 600 bombers, supported by 500 tankers and 200 fighters, and employed 160,000 people in the US and its bases around the world. By the mid-fifties America could field its B-52 Stratofortress which, along with the English Electric Canberra, is the longest-serving military aircraft in history.

Britain had its own nuclear force of V-bombers: the Vickers Valiant, the Avro Vulcan and the Handley-Page Victor. The Soviets squared up to Uncle Sam with the likes of the Ilyushin Beagle and Tupolevs Bosun and Badger.[8]

A US Air Force B-52 Stratofortress bombing
a target during the Vietnam war.

Overleaf:
Britain's V-Bombers flying in formation.
From left to right the Vulcan, Valiant and Victor.

At the height of the Cold War, there were always armed nuclear
weapons in the air on both sides, ready for immediate deployment. But the
nuclear stalemate that became known as MAD – for Mutually Assured
Destruction – meant that when the Cold War military aircraft actually went
into action it was in the way their Second World War predecessors had,
with bombers dropping 'iron bombs' (that is, conventional explosive
ordnance) from high altitude, reconnaissance aircraft spying on the enemy,
and fighters intercepting them and fighting each other. As much was seen
in Korea, Vietnam, the Soviet–Afghan war, the various Arab–Israeli wars
and the Falklands. All of Britain's V-bombers saw non-nuclear action –
the Valiant in the Suez crisis, the Victor during the Indonesia–Malaysia
confrontation, and the Vulcan in the bombing raid on Port Stanley.

In terms of what an aeroplane's role might be in achieving victory,
it seemed as though nothing much had changed since the end of the First
World War.

But soon enough, many of the world's military aircraft types would
begin to look completely redundant. The huge bomber fleets were
disbanded as the nuclear superpowers turned to missiles as a means of
balancing the threat of Armageddon. The future of big aeroplanes was in
carrying passengers.

The development of 'smart bombs', which could be launched miles
from the objective but could home in on a single building, rendered
theories about high-altitude and carpet bombing utterly obsolete. The future
of military aviation lay in smaller, multiple-role aeroplanes and helicopters.

The helicopter, a viable machine only after the end of the Second
World War, quickly became a brilliant battlefield device. The ominous
rhythmic *whump* of a Huey's rotors has become, in film and music, a

The Eurofighter Typhoon displays
its astonishing agility.

soundbite that defines our perception of the Vietnam conflict. Helicopters
not only offered brilliant operational versatility, they were excellent
weapons platforms as well, especially when equipped with sophisticated
radar that could automatically select and prioritise targets in seconds.
The helicopter became the Stuka of modern warfare.

At the close of the twentieth century, the military jet was no longer
in need of a qualifier such as 'bomber' or 'fighter'. It was all things a
warplane could be. The air forces of the world shrank accordingly, and
some nations were heading to the point where their front-line airpower
might be defined by only one aeroplane type.

The possible future direction of that type was demonstrated by the
Americans in the late 1970s, with their Lockheed F-117 Nighthawk. This
curiously angular aeroplane looked the way it did because it was designed

'THE TYPHOON'S
FLY-BY-WIRE
CONTROLS
ALLOW THE
AEROPLANE TO
BE INHERENTLY
UNSTABLE TO
THE POINT
WHERE, WITH
CONVENTIONAL
MECHANICAL
CONTROLS,
IT WOULD BE
UNFLYABLE'

to assume an attribute hitherto confined to popular mythology and achieved through magic: invisibility. A so-called 'stealth aircraft' is one with a low RCS, or 'radar cross section'.

The F-117 was developed by Lockheed's Advanced Project Development Team, the so-called Skunk Works, which had already produced several ground-breaking aeroplane designs (including the SR71 Blackbird). The angled surfaces of the F-117 were designed to minimise the reflection of radar signals. Its engine inlets were louvered, since conventional ones produce a distinct 'radar signature', and its infra-red profile was reduced by omitting afterburners, at the expense of rendering the aeroplane strictly sub-sonic. Its own navigation and weapons systems were 'passive' in that they did not emit any signals that might be detectable. The F-117 was strictly a bomber, with no defensive armament, designed to deliver its payload unnoticed. Even its bomb doors were designed to open and close in a few seconds, to minimise the time during which they were corrupting the carefully contrived shape of the thing.

Remarkably, this comic-book aircraft of the future, which saw action in Operation Desert Storm, is already being phased out, to be replaced by the Lockheed Martin F22 Raptor, an aircraft employing much stealth technology but capable of supersonic flight. Also in US service is the Northrop B-2 Spirit, a stealth 'flying wing' featuring a layout that has been the subject of much experimentation over almost the entire history of aviation.

Other, more recognisable combat aircraft are still with us, such as Russia's Sukhoi Su-35 and, more recently, the Eurofighter Typhoon. The Typhoon, the work of a European consortium from Britain, Germany, Italy and Spain, is a true multi-role aircraft. While it uses some stealth techniques, it avowedly relies more on its manoeuvrability and very sophisticated radar and weapons systems to defend itself. The Typhoon's fly-by-wire controls, in which extremely powerful computers interpret pilot inputs instead of acting directly on them, allow the aeroplane to be inherently unstable to the point where, with conventional mechanical controls, it would be unflyable. Instability yields superb agility; fly-by-wire and the lightning responses of the computer-operated flying surfaces mean it can be kept under control.

Remarkably, considering its performance and incredible responsiveness, the Typhoon has been described by one of its first RAF pilots as being 'very easy to fly', adding, 'as it bloody well should be at that price.' (A Typhoon costs in the region of £45 million.) Ease of operation and a highly intuitive cockpit layout are a deliberate ploy, intended to free the pilot to concentrate on his main task – which is not so much flying the aeroplane as deploying its weapons.

It is a philosophy that has taken some time to prove itself. It might be remembered that the attribute of being 'easy to fly' was once demanded of another front-line aeroplane, right back at the dawn of air power: the Royal Aircraft Factory BE2.

FULL METAL JACKET

An Israeli tank confronts a Palestinian taxi outside Bethlehem.

The 6-pounder gunners crouching on the floor, their backs against the engine cover, loaded their guns expectantly. We still kept on a zigzag course, threading the gaps between the lines of hastily dug trenches, and coming near the small protecting belt of wire, we turned left and the right gunner, peering through his narrow slit, made a sighting shot.

The shell burst some distance beyond the leading enemy tank. No reply came. A second shot boomed out, landing just to the right, but again no reply.

Suddenly, against our steel wall, a hurricane of hail pattered, and the interior was filled with myriads of sparks and flying splinters. Something rattled against the steel helmet of the driver sitting next to me and my face was stung with minute fragments of steel.

The crew flung themselves flat on the floor. The driver ducked his head and drove straight on. Above the roar of our engine could be heard the staccato rat-tat-tat-tat of machine-guns and another furious jet of bullets sprayed our steel side, the splinters clanging viciously against the engine cover . . .

The roar of our engine, the nerve-racking rat-tat-tat of our machine-guns blazing at the Bosche infantry, and the thunderous boom of the 6-pounders, all bottled up in that narrow space, filled our ears with tumult. Added to this we were half-stifled by the fumes of petrol and cordite.'
Lieutenant F. Mitchell MC, Tank Corp, France, 1918

The tank, more than any other single weapon or machine, has come to represent everything modern warfare is about. The image of the tank – triumphantly and incongruously parked in a town square or on the lawns of a government building, or advancing down a high street surrounded by

Right:
The clash of two new technologies.
A British tank under fire from a German
aircraft in the First World War.

Opposite top:
Little Willie, the first modern tank –
it was never to see action.

Opposite bottom:
Early tank crews wore 'chain mail'
to protect themselves against hot
flying metal.

ecstatic civilians, dust-caked and grinning liberators reclining on its flanks –
is one of the twentieth century's most iconic. The tank is like a self-contained
invasion unit, able to go anywhere and to destroy, or simply run over,
anything in its path.

The tank, like the aeroplane, first took centre stage in the First World
War, although the uses of the tank were rather more obvious and rather
better understood from the outset. H.G. Wells had set a precedent in his
short story 'The Land Ironclad', which appeared in the December 1903
edition of *Strand Magazine*. In this, Wells completely predicted the miserable
stalemate of trench warfare, in which two enemies faced each other across
a barren no-man's-land. But he also described the invention that would
break the deadlock: an armoured all-terrain vehicle that was immune to
bullets and could cross enemy trenches with ease.

Even then, the idea of an armoured advance wasn't entirely new.
Roman soldiers were known to organise themselves into small and tightly
knit formations while arranging their shields so that the whole formed a sort
of many-legged armadillo. Thus arranged, they could approach enemy
fortifications safe from the hail of missiles.

The sketches of Leonardo da Vinci, inevitably, show something that
looks rather like a flying saucer but which is, at least in concept, a tank of
sorts. The caterpillar track had been worked out in the late eighteenth
century and in the Crimean War of 1854–6 a small number of steam
tractors had been fitted with tracks to enable them to move around the
muddy battlefield.

But the idea of a heavily armoured motorised vehicle that could smash
its way through the enemy's lines remained little more than a curiosity until
1915, when Winston Churchill, then First Lord of the Admiralty, set up the
oddly named Landships Committee to investigate the potential of a decisive

new weapon.[9] And there was an urgent need for a new weapon, as by Christmas 1914 the First World War was already quite literally bogged down. Huge numbers of troops, on both sides, were trapped by mud and barbed wire.

The Landships Committee were hoping their own weapon would prove as effective as that of H.G. Wells and, being a committee, they set a number of criteria that the weapon would have to meet. It was to achieve a speed of 4 mph, climb a five-foot-high obstacle, and be able span a five-foot trench. What's more it should be armed with two machine-guns, and have an operational range of twenty miles.

In 1915, the first tank to step up to the mark was ready for testing.[10] It was given what is quite possibly the worst name ever conceived for a weapon, the mere mention of which was meant to strike fear into the hearts of its enemies: Little Willie. But Little Willie earned its place in history by being the forerunner of all modern tanks: it weighed around fourteen tons, was about twelve feet long and could carry three people in extreme intimacy. Sadly, it was not a great success. It failed in the requirement to cross a trench.

So a new tank was designed, this time with caterpillar tracks running around its whole body. Officially, it was called the Mark 1 but the urge to provide a stupid nickname could not be quelled and it became known as Mother. But at least Mother worked, and in trials at Hatfield Park in January 1916 it displayed its prowess by crossing a nine-foot trench. More incongruous tank nomenclature was devised: tanks with one or two large guns were known as 'male' tanks. Those with a number of lighter guns for infantry support were considered 'female'.

It was decided that the tank was now ready to be shown to Britain's leaders, the men who had the clout (and the cash) to decide its future. So, in great secrecy, Lord Kitchener, the secretary of state for war, and David Lloyd George, then minister for munitions, went to see Mother in action.

Kitchener was said to be less than impressed, describing tanks as 'mechanical toys' and stating that 'the war would never be won by such machines'. Thankfully for the tank, Lloyd George disagreed and ordered a hundred, to be delivered to France as soon as possible. Now all that was needed were some brave soldiers to drive them – not an easy recruiting job, because they were not to be told what they were expected to do.

The new tanks could not be readied as quickly as General Haig, commander of the British army in France, had hoped. They arrived too late for the big push at the battle of the Somme, where, on 1 July 1916, the British famously suffered their biggest day's losses of all time.

The progress of the war was becoming a desperate matter. Haig insisted that the tanks be ready for the autumn offensive, and the Mark Is, now all nicknamed Big Willie, were sent into battle near the villages of Flers and Courcelette on 15 September 1916. The tanks had a crew of eight and could rumble along at 4 mph. On the day, very few British soldiers had been told that the new weapon even existed, let alone that it was about to be deployed, so when the tanks clanked into no-man's-land, it was hard to know who was more surprised – the infantry they were meant to support or their enemy.

The tanks' first day of battle was not a great success, and Haig has since been much criticised for using them too soon. Of the thirty-six that saw action, ten were knocked out (some by the British artillery barrage) and fourteen either broke down or became stuck. Reliability would plague the early tanks for the rest of the war.

Furthermore, the conditions endured by the crews in action were discovered to be diabolical, as Lieutenant Mitchell recorded. The heat inside was tremendous and cordite fumes from the guns almost choked the occupants. The din of enemy fire striking a tank's hull was deafening, and when large-calibre rounds distorted the armour plating, shards of hot metal would break away from the inside face like shrapnel. This phenomenon, known as 'spalling', meant that the crew were forced to wear metal helmets ringed with chainmail, making them look like medieval knights (modern tanks feature an internal anti-spalling layer).

But if the tanks' effectiveness in France was open to question, it was not to be revealed back home. Newspapers heralded the tank as a new metal hero. They were described as sinister monsters spouting flame, and fanciful cartoons portrayed the Mark 1s as many-legged monsters rampaging across the battlefield. It was eagerly reported that the German army had fled in horror before them.

It wasn't long before the first pictures of tanks were published, in the *Daily Mirror*, which was said to have paid the enormous sum of £1,000 for them. The pictures were a huge boost to British morale and out of all proportion to what the tanks had actually achieved in France. Soon a film was produced showing the tanks in action. It opened in 1917 and millions of people saw it. Every time a tank appeared, audiences cheered and waved. The tank was an enormous success as a propaganda tool.

Left:
Big Willie's reputation in the British papers bore little relationship to its effectiveness in France.

Right:
An intimidating sight. The Mk IV appears over the top of a trench.

Hyacinth, a British tank from 'H' Battalion, gets bogged down in a German trench.

Overleaf:
The Russians are coming. The highly successful T-34 ready for battle in the Second World War.

Encouraged by the British experience, the French decided to build their own tanks, and in April 1917 sent 121 into combat in the Aisne offensive. Once again, the tank proved to be highly temperamental. France, though, was committed. Its lightweight Renault FT-17, which featured a fully rotating turret, would still be operational in 1940.

Despite the catalogue of disappointments, General Haig was convinced enough to order a further 1,000 tanks, and production was stepped up in Lincoln and Birmingham. The new tanks were ready for the third battle of Ypres, now remembered as Paschendaele. Once again, the tanks fared badly – the muddy ground was simply too soft for them. What's more, the Germans had recognised that the tank was able to cross their trenches. So they dug wider ones. Eventually, the British would respond with longer tanks.

Back in Britain, the tank was in desperate need of some more good publicity, so King George V was invited to visit the tank corp to see one for himself. Here he was treated to a remarkable demonstration; a tank climbed up one side of a six-foot-high bunker, and then crashed down the

other side, unblemished. After witnessing such a daring manoeuvre the King was concerned about the men inside, so he was quickly introduced to them. They looked well, and exchanged a few cheerful words with His Majesty. But he only met a few members of the crew. The rest were still inside, knocked unconscious by the ordeal.

At the battle of Cambrai on 20 November 1917, over 400 tanks were sent into action. They made an impressive start, breaking through a tangle of barbed wire, and in the first hour they had breached twelve miles of the German front, capturing thousands of soldiers, along with 120 guns and 280 machine-guns. Back in Britain, news of this success caused another outbreak of tank fever. Tanks were put on display all over the country and used to raise money for the war effort. A tank even starred in the Lord Mayor's Show. Poems and popular songs were written in celebration of the tank, and teapots were made in the shape of them.

In reality, the British did not have enough infantry troops to follow through at Cambrai, so, soon after the enormous initial success, the advantage was quickly lost to a German counter-attack. The outcome of the battle was to be considered a draw. Still, Cambrai restored the military's dwindling confidence in the tank. And for the first time even Germany began to build them. As they had captured fifty British tanks, they had plenty to learn from.

In the First World War, the German army was never entirely convinced by the tank; unsurprisingly, since they had had plenty of time to observe the shortcomings of the British ones. The German A7V was even less successful, struggling on anything other than level ground and requiring a crew of eighteen. By the end of the war, less than thirty had been produced, while Britain and France had built thousands. But the mere existence of the A7V meant that Britain's great war-winning innovation would now encounter a new obstacle: an enemy tank.

The first tank confrontation in history took place in April 1918. Three British Mark IVs – a development of the Mark I – took on three German A7Vs near the village of Villiers Bretonneux. Lieutenant Mitchell's account, reproduced above, is from this very encounter. He goes on:

Again we turned and proceeded at a slower pace; the left gunner, registering carefully, hit the ground right in front of the Jerry tank. I took a risk and stopped the tank for a moment.

The pause was justified; a carefully aimed shot hit the turret of the German tank, bringing it to a standstill. Another roar and yet another white puff at the front of the tank denoted a second hit! Peering with swollen eyes through his narrow slit, the elated gunner shouted words of triumph that were drowned by the roaring of the engine.

Then once more with great deliberation he aimed and hit for the third time. Through a loophole I saw the tank heel over to one side and then a door opened and out ran the crew. We had knocked the monster out.

German PzKpfw medium tanks invading Belgium, 1940.

Now the tank was ready for the final push to help bring an end to the war. They were deployed on a massive scale in August 1918, when 430 British tanks supported a twenty-mile advance against the Germans. It was to be the worst day of the war for the German army, but it came at a terrible cost in British men and machines. Of the 430 tanks that saw action on the first day, only 155 were still fighting the following day. By the fourth day only 38 tanks were still in service.

But the men and tanks had done enough, and now the German forces were retreating back to their home country, with the Allied forces on their tail. The war ended in November 1918, the new and newly proved tank spearheading the victors' final offensive.

But the tank would come back to haunt the Allies. Germany needed scapegoats for her military failure, and the tank, or perhaps Germany's want of them, now became one. Germany experienced a sort of inverse of the positive tank propaganda seen in Britain and, as a result, when the German army returned for another crack at Europe in 1939, it arrived in tanks.

The Allied tanks of the First World War had effectively spelt the end of trench warfare. In future wars, the front line would be mobile. Germany would prove this in 1939, when its well-drilled and highly mobile tank force squared up to France's static Maginot line. The Germans simply drove round it.

Between the wars certain basic truisms of tank design were embraced. Any tank was a compromise; a trade-off between weight of armour, speed across country and firepower. More armour meant lower speed, less armour meant less protection but better evasive performance. A big gun had to be mounted on a big tank, which then became a slow one. Three basic classes of tank evolved: light tanks, usually of ten tons or less and ideal for fast scouting; medium or cruiser tanks, intended for long-range operation; and main battle tanks, slow but heavily armoured and ideal for close infantry support.

It was realised that tank-on-tank confrontations were something to be avoided, since they were ultimately pointless. One job of light tanks, ironically, could be to forge ahead and disable the enemy's main battle tanks, so that they would not be available to hinder the bigger tanks of the advancing army.

But no one understood the value of tanks quite like the Wehrmacht. The invincibility of Germany's Second World War armour has been greatly exaggerated – in fact, its Panzer 1 was lightly protected and little more than a training machine. Even its much-vaunted Panzer IV weighed twenty-five tons and so was technically only a medium tank (the Germans were being as unimaginative as ever with their military nomenclature. 'Panzer' simply means 'armour').

But tank tactics made the German army invincible. In the Blitzkrieg, or 'lightning war', the German advance, headed by largely unopposed fast tanks, could move tens of miles in a day. Later, the Panzers would prove to be less than a match, technically, for the Allied armour. In particular, the Panzers performed badly against the Soviet T-34 during the invasion of Russia. The T-34 was in many ways the best tank of the war. It had better

Opposite:
An American M4 Sherman tank
in Tunisia, 1943.

Overleaf:
Israeli tanks advance towards Syria
during the 1967 Six-Day War.

armour than its German adversary, carried bigger armament, but gave nothing away in speed and manoeuvrability.

The effectiveness of the T-34's armour was largely down to its shape. Its plating was carefully arranged to present a sloping face to an enemy shell, which would often simply be deflected before it could penetrate or explode. German tank crews were reported to be baffled when they saw direct hits on T-34s simply bounce off harmlessly. Sloped armour became a trademark of post-war tank design.

Other radical ideas proved less convincing, most notably the DD tanks used by the Allies during the Normandy invasion. The DD tanks, mainly American Shermans, were amphibious, and DD stood for duplex drive: power could be directed to the tracks in the usual way or to a propeller. Buoyancy, after a fashion, was achieved with a flotation skirt. (It wasn't a completely new idea. Amphibious trucks had already been used with some success.) The DDs were designed to provide support to the first troops to reach the beaches, as it was known they would come under heavy machine-gun fire from the Germans' beach-head bunkers. The DDs would be launched two miles offshore to make their own way to the beach ahead of the infantry in their cumbersome landing craft.

Amphibious tanks achieved mixed success during the landings. At Sword beach, where the British landed, the sea was reasonably calm and thirty-two of the thirty-four DD tanks made it to the beach. The results at Gold beach were less impressive. Rough seas forced the landing-craft commanders to drop the tanks right at the water's edge, which meant that they did not arrive ahead of the infantry and their advantage was lost.

But the DDs performed worst during the American attack on Omaha beach, where twenty-seven of the twenty-nine DD Shermans launched sank. There has been much debate as to why this happened. The rough water didn't help – the DDs were designed to cope with waves a foot high, but at Omaha encountered waves of up to six feet. The tanks were also released too far out, about three miles offshore. Many of the crews were drowned, and some of the DDs remain, to this day, in the waters off the French coast.

The course of the war also served to undermine some of the peacetime thinking on tank classification. Tanks would, and did, fight extensively among themselves, notably in Africa, Russia and during the invasion of Europe. So the light tank fell out of favour as the inevitable armour and firepower escalation rendered them too weak, their scouting role becoming the work of armoured personnel carriers and half-tracks. Germany's Panzer IV weighed twenty-five tons and was fitted with a low-velocity 75 mm gun. The later Panther boasted a bigger and more destructive high-velocity 75 mm gun and weighed forty-five tons, but was seen as something of a medium tank next to the mighty Tiger, which weighed fifty-seven tons and was armed with a much-feared 88 mm gun. As long as tanks encountered tanks, the biggest tank seemed to be the best. Britain ended the war with the 38.5-ton Churchill and America with the 42-ton Pershing.

Another problem for the tank was the development of effective anti-tank weapons, some of them incredibly discreet and mobile, such as the

'BY THE END OF THE SECOND WORLD WAR THE PATTERN OF TANK DEPLOYMENT AND THE BASIC REQUIREMENTS OF SUCCESSFUL TANK DESIGN HAD BEEN WORKED OUT TO THE POINT WHERE THEY WOULD NOT CHANGE SIGNIFICANTLY'

British PIAT (a name derived, in typical British army fashion, from Projector, Infantry, Anti-Tank), the American Bazooka and the German Panzerschreck. It is because of weapons like the PIAT that tanks were gradually fitted with deep side-skirts, there to protect as much of the exposed and relatively fragile tracks as possible. These gave rise to a great paradox of tank operation. A tank's crew enjoys the greatest personal protection when all the hatches are shut. But in this configuration, the tank is vulnerable to sneak attack, since visibility is greatly reduced. It is for this reason that tanks are usually seen, even in the heat of battle, with the main turret hatch open and the commander standing up in full view.

By the end of the Second World War the pattern of tank deployment and the basic requirements of successful tank design had been worked out to the point where they would not change significantly. Tanks of the Cold War, such as Britain's Centurion, Germany's Leopard and the Soviet T-54, were the same in appearance as those that had ended the Second World War. Tank battles would continue, for example in the Yom Kippur war, and tanks would operate as an advanced invading force, Blitzkrieg-style, in Desert Storm.

US Marines, armed with a Bazooka, wait for their moment to open fire during the Korean War.

One brave anti-government demonstrator challenges a column of Chinese Type 59s in Tiananmen Square, Beijing, 5 June 1989.

Development was concentrated instead on the tank's operating systems. The defeat of Rommel in the desert, partly through lack of fuel, led to the development of multi-fuel engines that would run on petrol, diesel, or any volatile fluid in between. Gyroscopic devices kept the gun level even when the tank was traversing hilly terrain. Simple reticular gunsights gave way to accurate optical devices, and by the end of the century the satellite navigation that was used to guide the tank was also being used to guide its shells. Tanks used radar, and attempted to hide from enemy radar through stealth techniques such as lowering the thermal signature of the engine's exhaust. The crew compartments of Cold War tanks such as Britain's Chieftain could be sealed against chemical attack, and the crew could live inside, in a level of discomfort that would have appalled even Lieutenant Mitchell, for weeks.

But a tank was still a tank, a mobile big gun, designed for invincibility in the face of anything other than another tank, which was a situation to be avoided. Perhaps the most poignant tank image of the late twentieth century is the photograph of a lone protester, later known simply as 'the tank man', confronting a line of Chinese Type 59s during the Tiananmen Square protests of 1989. Although he was named by *Time Magazine* as one of the hundred most influential people of the century, he failed to stop the advance.

As in the trenches of 1916 and the French villages of 1939, the tank is at its most effective when deployed against people who don't have them.

TOP SECRET

Descriptive Catalogue

OF

Special Devices

AND

Supplies

COMPILED & ISSUED

BY

M.O.1. (S.P.)

THE WAR OFFICE.

1944.

I SHALL SAY THIS ONLY ONCE

'THE SOE CATALOGUE WAS A GEM, AN ARGOS CATALOGUE OF THE UNDERWORLD'

It is commonly believed that military satellites floating in space can read a newspaper on the ground. In fact, it turns out that the best resolution available through satellite imaging is a blob about the size of a large tablecloth.[11] Allegedly. So you can sit on a bench in New York's Central Park with a copy of *Das Kapital* without worrying that Satellite McCarthy is reading over your shoulder.

Or can you? That might simply be what they want you to think. The trouble with any attempt to understand the hidden world of espionage is that the important bits are still secret. Of course, reformed spies write memoirs, but by the time they are published the spy community has moved on. And how do we know that the memoirs of Agent X aren't just a feint anyway, or that we're not just living in a massively complex Alistair MacLean plot?

Espionage is, like armaments and computer software, a technology race, with the initiative in constant flux. Once, it was a gentleman's profession in which people wearing absurd powdered wigs hid behind the floor-length curtains of a foreign embassy, or lurked in bushes with a brass spyglass. But in the twentieth century, the era of technology, it became a mainstream profession in which spies benefit from salaries, career progression, annual holidays, company cars and sick leave. Today's MI6 spooks (they prefer to be called 'intelligence officers') appear on Radio One exhorting the young of Britain to join up and enjoy glamorous foreign travel. But it could just be a front.

The job does have its drawbacks, such as torture, imprisonment without trial and sudden death in the form of an innocent-looking piece of sushi. On the other hand, there are some great gadgets, and not just invisible ink made with lemon juice.

All schoolboys would secretly like a Walther PPK pistol.[12] The genuine article was originally designed for German plain-clothes policemen because it was easy to conceal, but rose to fame as James Bond's weapon of choice. Or they might fancy the silenced 7.62 mm Tokarev TT-33, once favoured by the

The cover of the 1944 SOE Catalogue.

Soviet counter-intelligence agency SMERSH. (Firearms experts object to this word. What we call a silencer is actually a suppressor.)

Women might be more interested in the lipstick pistol, a gun that looked exactly like a tube of lipstick and could even be used as one, given care. But the bottom portion of the case unscrews to reveal a chamber for a single round, and the base of the tube is fitted with a powerful spring and firing pin which, if pulled and released sharply, fires the bullet. It was once issued to female Russian agents. It was, of course, designed to be used at very close range and was really a means of knocking out an enemy agent and then taking his or her weapon. As well as death, the victim suffered the added ignominy of being splattered with tiny flecks of the popular Soviet cosmetic Red Death, but that would probably be hushed up.

The lipstick pistol actually comes from a fine tradition of concealed weapons that fire just one bullet. There are many other ingenious one-shot pistols disguised as a smoker's pipe, a propelling pencil, or a torch. But if disguising the weapon wasn't going to fool the enemy, then it could always be hidden. The KGB developed a single-shot gun designed to be concealed in your rectum. (It wasn't designed to be fired from your rectum. You were supposed to remove it first.) Strangely, we never saw Q demonstrating this.

Such was the demand for secret weapons and clandestine equipment during the Second World War that the stuff had to be manufactured on a grand scale but, at the same time, in great secrecy. Consequently, the British Special Operations Executive, who were responsible for sabotage, subversion and intelligence in Nazi-occupied Europe, operated their own workshops and recruited a variety of skilled military and civilian technicians, many of whom had previously worked in the film industry. From June 1942, their principal location was the Thatched Barn on the Barnet Bypass, north London, code name Station XV. But, unusually for an organisation that shunned detection, the SOE also had a shop window and a catalogue.

The venue for the 'shop' was behind the scenes at the Natural History Museum in South Kensington, unbeknown to hordes of visiting civilian dinosaur enthusiasts. Here a massive range of secret equipment went on display in order that agents could familiarise themselves with the latest gadgetry. The gallery also proved valuable for public relations; even King George VI paid a visit. One VIP recorded in his diary: 'Friday 10th November. The South Kensington Natural History museum, where in six sealed rooms they have an exhibition of all gadgets. Very ingenious containers and one-man submarines, also many kinds of disguised bombs and explosives which stick to ships. Wonderful dummies, too, with dreadful faces and dressed in the different German uniforms. A good show'.

The SOE catalogue (or, more precisely, the *Descriptive Catalogue of Special Devices and Supplies*) was a gem. Its contents ranged from simple entries such as 'Cable, Electric, Duratwinflex. Resistance is 4.72 ohms per hundred yards, double' to the highly technical 'Microwave Equipment Aircraft Installation, Type A2'. The catalogue had sections on Small Boats,

 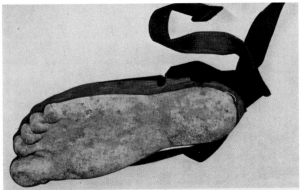

With 'sneakers' you could enjoy army boot comfort while making the right impression.

Camouflage, Incendiary Stores and what were called Ad Hoc devices. It was beautifully illustrated, an Argos catalogue of the underworld. Many of the items were breathtakingly devious: explosive coal, explosive soap and explosive rats. Also listed were boots, canvas, anti-mud; these could be worn over normal shoes when covering rough terrain only to be removed once the spy had made it to firm ground. And how about Sneakers? Not the trainers we know now, but a type of footwear on which the sole was imprinted with the shape of a Japanese shoe or a native foot, thus allowing intruders to cross sandy or muddy areas without leaving a telltale trail of prints from standard-issue army boots.

Elsewhere, skilled artisans were busy creating escape equipment that could be hidden in parcels sent to the inmates of Colditz and the like. Maps, small tools, money, false papers, and of course, compasses were hidden in everything from dominoes to shirt sleeves. Anyone who grew up in the 1960s might remember the homage that was paid to these people's work by the shoemaker Clarks, whose Commando shoe featured a small compass hidden in one heel – ideal for anyone who needed to escape from geography.

The SOE followed Winston Churchill's orders to 'set Europe ablaze' and played a vital role in winning the Second World War, but once it was over the whole operation was disbanded. In the Cold War, sabotage would give way to espionage, a more clandestine battle in which information was actually the weapon that kept Armaggedon at bay, not nuclear warheads as we all thought.

This was the era of MICE. MICE was a typically absurd secret service acronym denoting the four basic routes through which a spy might be recruited from the other side: Money, Ideology, Compromise and Ego. These things had long been stumbling blocks for anyone trying to keep a secret and useful tools for anyone trying to find out what it was, but now MICE was used at a governmental level for infiltrating the enemy's psyche.

Money was a great way to turn a mind, especially if the target was in deep debt. Once money had been accepted, it was difficult to back out of an arrangement without humiliating exposure. In the US Aldrich Ames was arrested in 1994 having worked as a KGB mole in the CIA. He was paid over $2.5 million. He said the money was to pay off his debts and his wife's shopping habit.

Rather less tawdry was ideology. The famous Cambridge spies Anthony Burgess and Donald MacLean were determined that Communism was the way forward, and so were persuaded to feed information to the Soviet Union from their high positions within the British Secret Service.

Compromise was once easy, especially when homosexuality or mere kinkiness were enough to put you at risk. The KGB used to keep rooms in hotels for the purposes of drawing unsuspecting Westerners into compromising positions, and then photographing them.

Finally, there is ego. Naive academics working for a government, for example, can be flattered into revealing much more than they should.

Still, once brow-beaten into becoming a spy, there was at least some great kit to play with, most notably cameras of all sizes and outward appearances. Cameras were concealed in umbrellas, in matchboxes and in cigarette lighters. And these are just the ones we know about. A more cunning device was a camera that really was just a camera, but fitted with a lens cap that appeared opaque but allowed photographs to be taken straight through it, which was great for deceiving border guards. There were also beautifully made miniature tape recorders, special keys for unpicking locks and, of course, a huge range of bugs.

A cunning British Second World War device for disabling the German railway system by means of rodents, exploding.

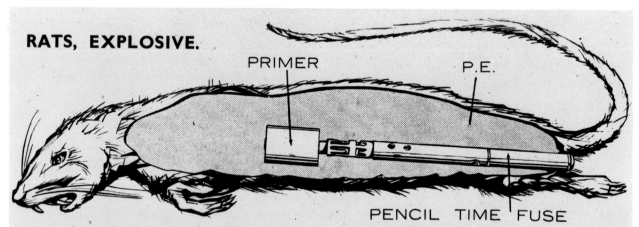

RATS, EXPLOSIVE.

PRIMER

P.E.

PENCIL TIME FUSE

A rat is skinned, the skin being sewn up and filled with P.E. to assume the shape of a dead rat. A Standard No. 6 Primer is set in the P.E. Initiation is by means of a short length of safety fuse with a No. 27 detonator crimped on one end, and a copper tube igniter on the other end, or, as in the case of the illustration above, a P.T.F. with a No. 27 detonator attached. The rat is then left amongst the coal beside a boiler and the flames initiate the safety fuze when the rat is thrown on to the fire, or as in the case of the P.T.F. a Time Delay is used.

'LESS OBVIOUS
IS A TRULY
MINUTE CAMERA
INCORPORATED
INTO THE KNOT
OF A SPECIAL
MI6 TIE SUPPLIED
BY SPY RACK'

The successor to our bewigged man behind the curtains became a remarkable device, especially in the age of true miniaturisation. A demonstration by an ex-MI6 man, who claimed to be able to kill with a fork (this may have been a bluff. It could have been a spoon) demonstrated just how easy the art of basic bugging has become. In a re-creation of a common initial recruitment test, a simple and innocent-looking office is multiply bugged, the challenge being to see how many the prospective spook can find. Some are fairly obvious, such as a small camera hidden in the spine of a book, and especially if the bugger has been daft enough to use a conspicuous book that appears to have no place on the shelf (in this case a volume on Pilates 'hidden' on a shelf full of war stories). Similarly, a small camera positioned in the leaves of a pot plant is easy to spot once you look for it. Less obvious is a truly minute camera incorporated into the knot of a special MI6 tie supplied by Spy Rack.

To find cameras more cleverly incorporated into, say, wall clocks and picture frames demands a bit of counter-surveillance technology in this case a small eyepiece equipped with flashing red LEDs. When viewed through this, the curved lenses of cameras produce tiny red reflections and reveal themselves. But in doing this on film you will have revealed that you're expecting to be watched and therefore have something to hide.

Radio-based listening devices can also be detected, but only when they're running. A more discreet type of audio bug, in this case hidden in a desk lamp, only works when activated remotely by a mobile telephone. The rest of the time it is dormant and undetectable. Most alarmingly, this lamp and everything described above is available on the high street, which leads to two obvious questions: how many voyeurs have this stuff, and how good is the kit we don't know about yet?

It's tempting to think that spying was more fun in earlier low-tech times, when, for example, Soviet agents operating in London were advised to use expensive luggage when checking into the Dorchester or Ritz, because the British tend to spend a lot on smart cases. But even then there were spies, counter-spies, and counter-counter-spies, and the world of espionage has probably always functioned in a state of near stalemate.

Consider the Clarks Commando shoe again. Many schoolboys in the sixties and early seventies would carry a very evident Swiss Army penknife or home-made ballpoint blowpipe, knowing that teachers would confiscate these things and never think to look for the compass. As a result, thousands of them made successful home runs every day. Only to be recaptured the following morning, having failed to realise that their parents were double agents.

BODILY FUNCTIONS
OUTSIDE
INSIDE

6

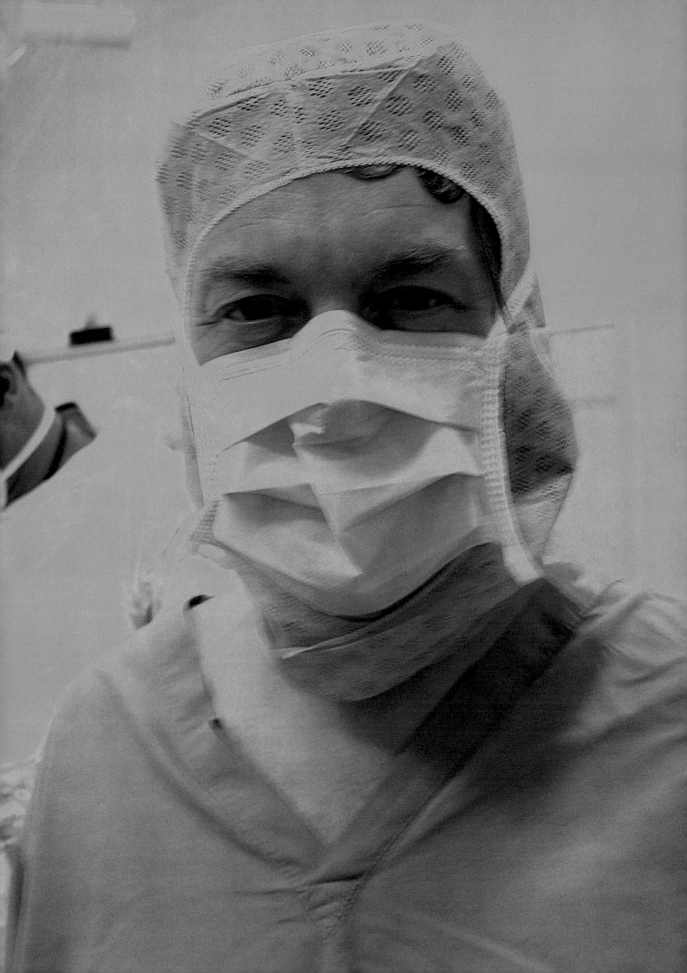

THE RIGHT TROUSERS

'Buzz' Aldrin models the A7L lunar suit.

For thousands of years, humankind's innate desire to improve its lot was really hampered only by the limits of courage and curiosity. When primitive man ventured beyond the next hill or the next valley to increase the scope of his hunting and gathering, the threats were largely being eaten by other animals and falling down holes of one form or another. But like all mammals, the human species has evolved such that a healthy example should always survive an impact with a solid object when travelling at its maximum speed.

Admittedly, some extreme forms of exploration could reveal a certain frailty of the human form. Burke and Wills, lost in the Australian interior, succumbed to the extreme heat, not being natives of the land and therefore not fully equipped to deal with it. And the British, a nation of explorers with notoriously poor dress sense, have several times demonstrated the folly of venturing into the extreme cold in inappropriate clothing. On the whole, though, the limitations of our bodies went unchallenged.

All this began to change with the machine age, and especially once the twentieth century got under way. The depths of the oceans, the thin air of the stratosphere and the vacuum of space became the new frontiers of exploration and recreation, and they could not be conquered or enjoyed without some form of life support.

Before all that, though, there was the issue of speed. This had been on the increase since the industrial revolution and the invention of railways, which propelled their passengers at such huge velocities – sometimes up to 20 mph – that many feared they would suffocate. More precisely, the real issue became *acceleration*, which produces the phenomenon we usually call G-force but which proper physicists refer to simply as *g*. It is an accelerative force equal to the pull of gravity, and roughly equal to $9.8m/s^2$. It is not simply felt when travelling in a straight line but also when describing any circular path, because any body travelling through a curve is always accelerating towards the centre of its radius.[1] This explains why you might want to throw up on the waltzer.

Anti-*g* trousers in deflated mode.

Right:
Me, in deflated mood. The Siebe
Gorman diving suit is possibly the
world's least comfortable ensemble.

Overleaf:
A modern Scuba diver on holiday,
and having a much better time.

A Porsche 911 that can accelerate from 0 to 60 mph in 4.1 seconds will subject its driver and passengers to a force of 0.7 *g*. The acceleration of a Formula One car can produce 3 *g*, a roller coaster perhaps 4 *g* (though only for a second or two) and the launch of a space rocket, 5 *g*. A modern combat aircraft can easily create 9 *g* and even up to 12 *g* in extreme manoeuvres. In 1954 Colonel John Stapp, while conducting a series of experiments on the effects of extreme acceleration on the human body, is said to have sustained over 45 *g* on a rocket sled.

So as soon as cars started racing around banked circuits and aeroplanes started making tight turns, *g* became a problem. Just how debilitating this force is can be experienced on a centrifuge, for example the one that has been part of the Farnborough Institute of Aviation Medicine since the mid-1950s. Looking like a giant fairground attraction with its long rotating arms and pivoting capsules, it is actually a device for training astronauts and pilots and for testing their equipment.

For a normal adult (me, in this experiment) 3 *g* is reasonably tolerable, although it means the victim's head and arms weigh three times what they did when stationary. Increasing the force to 4 *g* causes greying of the vision and adds about 40 years to the face. At 4.4 *g* I finally suffered what is known as *g*LOC, *g*-induced Loss Of Consciousness. Once the sadistic doctor had stopped the ride I came round to talk gibberish in complete safety, but had I been flying an aeroplane the results could have been disastrous.

Reports of *g*LOC are almost as old as the aeroplane itself: it is believed that one pilot suffered as much in a Sopwith Camel in 1917. Pilots in the Schneider Trophy air races of the twenties also reported 'greyouts'. But the problem reached a head with the development of the Second World War's high-performance fighters and, more notably, in the fast jets that followed. Suddenly, the performance of the machine was being limited by the performance of its operators. Turns out they were wearing the wrong trousers.

The first anti-*g* flying suit was created in 1941 by a team at Toronto University led by Wilbur Franks. The legs of his anti-*g* trousers incorporated water-filled bladders, and they were intended for Hurricane and Spitfire pilots. American pilots also tried Franks's suit but found his water system uncomfortable, so were issued instead with an air-inflated version known as the Berger Suit.

Whichever tailoring you preferred, the principle was the same. The trousers, when they were filled with either water or air, squeezed blood out of the legs and kept it in the brain, where it was needed.[2] Back in the centrifuge, the sensation is unusual, and trousers as tight as these have not been experienced by civilians since the 1970s. But they do work, and 5.4 *g* was easily tolerable, thanks to nothing more than a timely squeeze of the thigh from a compressed air system in the 'cockpit' that senses increasing *g* and acts accordingly. Passengers in today's high-speed military jets, such as the Eurofighter Typhoon, can relax in an all-enveloping anti-*g* suit and wave goodbye to *g*-induced blackout misery completely.

So in the air it became possible to add a modification that allowed the body to exceed its natural limits. Water, though – or under it, at least – is a truly hostile environment for humans and one where mere survival is impossible without help. The odd pearl diver is able to hold his breath for minutes at a time, but for most of us one underwater length of a swimming pool is a near-death experience.

Diving in a suit supplied with air is actually quite an old idea, and the standard deep-sea-diver outfit known as the Siebe Gorman suit dates back to the first half of the nineteenth century (the helmet part of the design never changed, and remained in production from 1837 to 1975). The actual suit is a hideously heavy, claustrophobic and smelly ensemble, and not something that would be worn unless absolutely necessary. Necessary it was for the construction of some of our great bridges and in the pursuit of one or two famous salvage operations (like the recovery of fifty-nine bars of bullion from the wreck of the SS *Skyro*, which sank in twenty fathoms off Finistere in 1891). Similar suits are still in use for really serious and deep diving. Diving for a hobby, though, required something less intimidating.

The two people who brought us the Aqualung were Jacques Cousteau and Emile Gagnan in 1943. It was a successful form of Self Contained Underwater Breathing Apparatus, or SCUBA, which did away with umbilical links to the surface and allowed the diver to carry a supply of compressed air in a bottle. (There had been earlier attempts at self-contained breathing apparatus, but none terribly successful and one or two positively fatal.) The most important part of Cousteau and Gagnan's innovation was the demand valve, which does two jobs: it reduces the high-pressure air in the air cylinder to ambient pressure, so that when you breathe in you don't

Inspector Clouseau as Jacques Cousteau, or vice versa, preparing to enter the murky depths.

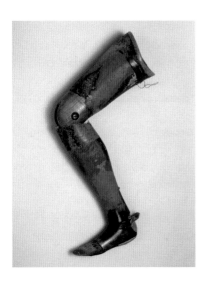

An unusually sopisticated wooden leg made in 1894 for a 16-year-old soldier. He wore it for sixty years.

blow your head off, and it ensures that the air only flows when you actually suck on the mouthpiece, so none is wasted.

This may not sound like much, but it means that ordinary people can gaze at brightly coloured fish by paying for an expensive diving holiday, instead of having to pay for an expensive dentist and then admiring his tropical aquarium. The social impact of the aqualung should not be underestimated: in the US alone around half a million new SCUBA divers are trained every year.

The technologies that allowed people to survive the perils of high-speed flight and explore the airless ocean would obviously have some influence on space attire. The Model A7L suit worn by Neil Armstrong and Buzz Aldrin on the moon was designed to protect the two astronauts against a multitude of cosmic hazards. Along with its primary life-support system, which carried the air and power supplies, the suit weighed about 81 kg. But on the moon, thanks to its reduced gravity, its weight was a far more manageable 13.6 kg. Putting the suit on involved clambering in through a rear pressure-sealing zip that ran from crotch to the upper back. The suit was built up from 25 layers of protective materials to survive being struck by micrometeoroids and extreme temperatures ranging from −150 °C to +120 °C. In the suit the astronaut needed to be able to walk over the harsh lunar terrain, and, if he should tumble, pull himself up unaided.

Adding extra functionality to the bodies we were born with is all very well if the aim is extending the boundaries of exploration and experience. But the other great contribution of the twentieth century to the human body was in replacing original bits that had been lost.

During the First World War, anyone who lost a leg would expect to be provided with little more than a strap-on broom handle and a walking stick, and would do well to walk unaided. Later, more convincing artificial limbs would be made with sheet aluminium or wooden carvings, but were still nothing more than representations of an arm or a leg fitted with a rudimentary hinge. The emphasis was largely on disguising the loss.

But in the latter half of the century the thinking of prosthetists changed. They began to wonder if maybe their job was not to produce something that looked like an arm or a leg, but rather something that worked like one. They stopped thinking of themselves as panel-beaters and carpenters, and became engineers. The result is that a modern artificial leg looks less like a leg than one of the post-war era, but does a better job of being one. A below-the-knee amputee of today can be fitted with a powered replacement leg which includes a small stepper motor in the ankle joint to provide an impulse to each step, thus replicating the action of the original instead of simply hanging passively. The only requirement of the wearer is the need to recharge the unit every few days. Experiments with linking these motors to nerve endings is seen by some prosthetists as 'the first step towards bionics'.

The development of simple and quick plug-and-socket systems for the attachment of new limbs also led to another radical idea – why just one arm or leg? Why not a range to suit different activities? Today's paralympians

An artificial leg from around 1900.

Opposite:
Britain's John McFall in the hundred metres, 2007.

will often own several prosthetic arms and legs for different sports. One athlete interviewed in the research for this book, an above-the-knee amputee, owned four different legs for sport, plus the old-style wooden one he had been given in his youth. That gave him six legs, including his surviving natural one, which made him, as he put it, 'technically an insect'.

Paralympians, in fact, have become the source of some controversy in athletics circles, especially now that it is accepted for them to compete against the 'able-bodied'. Complaints have been made to governing authorities to the effect that amputees running on the likes of the Cheetah leg have an unfair advantage. It's a tricky debate, but a great compliment to the work of the new-era prosthetics engineers.

Next, perhaps, someone will perfect the concept of an artificial exoskeleton to improve body strength, as seen in the closing scenes of the film *Aliens,* when Sigourney Weaver puts on her powered metal suit to give the alien a good kicking. Needless to say this is an idea the US army have been toying with for some time: for combat, faster mobility and for carrying heavy loads, but so far the exoskeleton remains the stuff of science fiction. Except, that is, for one maverick Canadian inventor, Troy Hurtubise, who has made the development of an exoskeleton his life's work. His first major invention, a 'grizzly suit', was designed to protect the wearer against attacks when observing bears at close range. Youtube enthusiasts may have seen the hugely popular clips of Hurtubise throwing himself off precipices and being struck at high speed by pick-up trucks, without injury.

Hurtubise has been widely parodied, but are his ideas really that daft? No dafter, surely, than trousers that keep you conscious, a mask for breathing underwater, a suit for moon walking and artificial legs that can outsprint natural ones. It was all unthinkable just a lifetime ago.

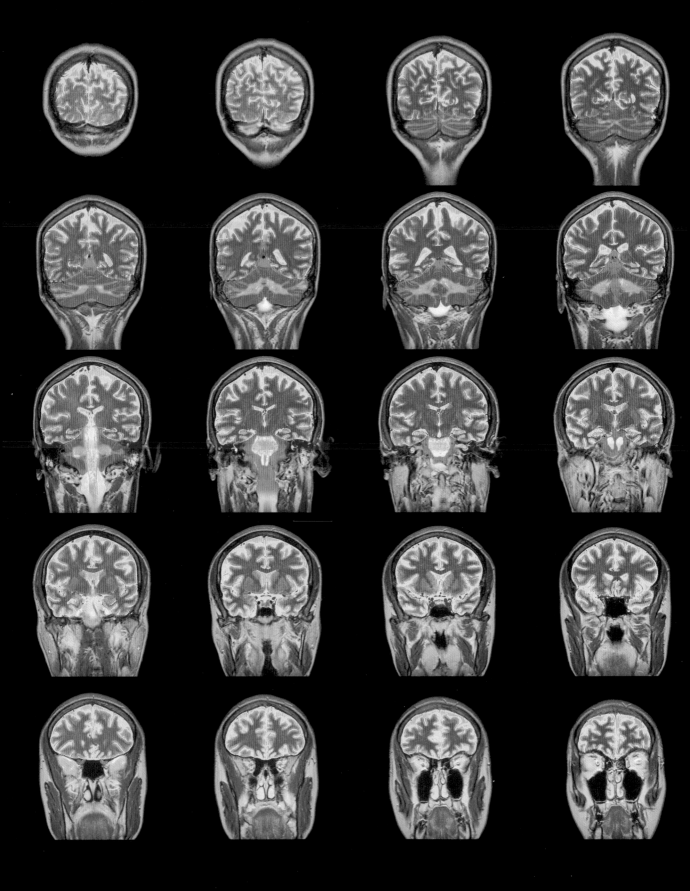

CONFRONTING THE INNER SELF

MRI scans through the skull
of a 31 year old woman.

The humble apple, it transpires, is a complex proposition. The shape is simple enough, but then there's the colouring, the texture, the flavour, the sensation of your incisors piercing the firm skin, the smell of the thing under your nose. Merely contemplating the mystery and wonder of the apple, the fruit with which the artist Cezanne promised to 'astonish Paris', involves quite a bit of brain activity.

This much can be revealed with a Magnetic Resonance Imaging (MRI) scan of the brain.[3]

The MRI machine subjects the cells in the grey matter to a pulsing magnetic field, which has the effect of magnetising them and aligning them. As the field collapses, the cells return to their regular random arrangement, but some more quickly than others. By detecting how long it takes for the cells to 'relax', a three-dimensional map of the brain can be assembled by a powerful computer and presented as an image. It is not a photograph as such, but a complex record of molecular activity displayed in graphic form. To all intents and purposes, though, it produces 'a picture of your brain'.

That is a gross simplification of the science, which is so complicated that its discovery was the stuff of several Nobel Prizes reaching back to 1944. It was not until 1977, however, that the technique was used as the basis of a scanning device. The first to actually create an image of a human being was the American Raymond Damadian, who named his machine the Indomitable. It is now preserved in the Smithsonian Institute in Washington. And rightly so, because it was to allow non-invasive study of the most elusive organ in the human body, something that had not been possible with mere X-rays. More controversially, the MRI scanner can record brain activity.

Back to the apple. The scanner employs a magnetic field around 60,000 times more powerful than the earth's, so before I could be inserted into its (slightly claustrophobic) tunnel, I had to be checked for metallic content. This is essential to avoid any giant-magnet comedy moments that

'THE MRI MACHINE ESTABLISHED THAT I HAVE A VERY BIG BRAIN'

belong in a cartoon rather than a serious medical research centre. Metal pens in pockets and buckles on belts obviously have to be left outside, but I also had to be checked for old shrapnel wounds, pinned bones, metal teeth (normal fillings aren't a problem) and hairpieces with ferrous fittings.

Once in the tube, I was shown a series of pictures and asked to think about them very hard. They included the apple, some other fruit, some emotionally rather neutral items such as a dining fork, and a few things I am known to like, such as the Aston Martin V8 Vantage (the proper 1975 model, not the poncy new one). As my brain pondered the relative merits of cutlery and cars, my neural activity was being tracked, as the scanner could detect the resulting rush of blood and oxygen to various bits of the brain. All of this can be displayed on the resulting series of images and colour-coded for ease of interpretation. (Strictly this machine was an fMRI scanner, the 'f' for functional, a version of the technology specifically intended for brain work.) First, the MRI machine established that I have a very big brain, although medical science suggests there is no clear correlation between this and intelligence. (But that might just be a rumour put about by people with small brains – see the bald-men-are-more-virile analogy.) More remarkable was the activity in the amygdala, areas of the brain associated with emotional responses. It is because they were 'lit up' during my most sinful apple thoughts that we know how complex an apple is. The picture of the fork generated nothing like as much neural activity, despite thoughts of chips.

The picture of the Aston yielded even more: not just a great deal of emotional brain baggage, but stimulation of the motor cortex, the part of the brain controlling practical activity. Because I was thinking about the act of driving a car, it came out in red on the scan. I now have on my wall a print-out of this image, a self-portrait of my brain with thoughts of Aston Martin ownership. Cezanne didn't do anything quite so astonishing.

So does MRI have the potential to form the basis of a mind-reading technique, and could the scanner become the Tube of Truth? No, say its advocates. It can never read thoughts, it can only show that thought is going on. Neither does an MRI scanner provide any cures; rather, it is an analytical tool that will spot problems with the brain long before they manifest themselves elsewhere, thus making a cure more likely and, hopefully, simpler.

The twentieth century finally allowed the body to be explored without the horror of surgery. X-rays had been known in the nineteenth century, and it was in 1895 that Wilhelm Röntgen demonstrated that they could be used to reveal bone structures.[4] But they have never been so good for exploring soft tissue, except where the anomaly is sizeable, such as a lung problem. They have, however, been used to produce CAT (Computed Axial Tomography) scans, developed in 1968, which provide a three-dimensional image of skeletal structure.

In the late 1950s Professor Ian Donald reasoned that if sonar could be used to find a submarine, something similar could be used to see beneath the surface of the body. His work gave us the ultrasound scan, familiar to pregnant women as the safe way to check the health of the foetus.

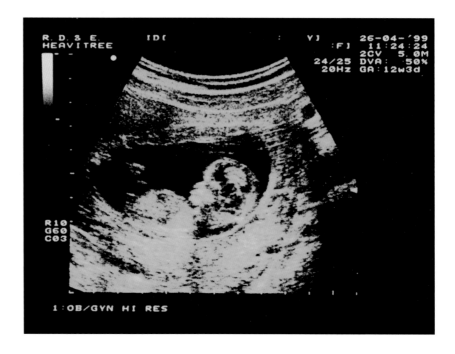

Other forms of ultrasound scanning are used to diagnose problems with the testes, kidneys and, most notably, the heart.

In the early years of the twentieth century the guidelines referring to operations on the heart hadn't changed much for thousands of years. They were, in a nutshell, don't meddle (in modern heart operations the organ can be stopped, but the patient is not technically dead). Not only is the heart, through its pulse, the denoter of life itself, it is deeply buried, and inaccessible without subjecting the patient to massive trauma. Fortunately, one or two pioneers chose to confront the problem.

One man in particular, Werner Forssmann, deserves a special mention. In 1929, Forssmann, then at Berlin University, had seen how a very long, very narrow rubber catheter could be inserted in a vein, in this case the vein of a horse. Forssmann believed that if he could master this technique on humans, he might be able to use it to deliver drugs precisely to parts of the body that are otherwise very hard to reach, such as the heart. He tried his idea out on a corpse, and it worked. Now he needed a living guinea pig. So he volunteered himself.

With the help of a colleague he pushed a catheter into a vein at his elbow. The catheter was more than two feet long, so Forssmann estimated it would reach his heart. The colleague who was assisting him became nervous and left Forssmann to it, but Forssmann continued. He roped in a nurse to help him, and continued to push the catheter towards his heart while she held up a mirror to an X-ray machine. In the reflection he could make out the position of the catheter deep inside his vein. Then, he reported feeling a burning sensation and he could see that he had reached the right atrium of his own heart. Then, with the presence of mind to record the

This may look like part of a washing machine, but it's the first artificial heart to be fitted to a human being, the Jarvik 7.

event, he went upstairs to find someone who could take a photograph of his achievement. Finally, he lay down, dazed, while a rumour spread around the hospital that he had committed suicide. But he lived, and some years later was to be awarded the Nobel Prize for his valiant work.

Forssmann had shown that the heart could be reached, and possibly cured of ailments, without the knife and the hideous powered buzz-saw used to split the rib-cage in an open-heart operation. But if surgery became unavoidable, then it would help immeasurably if the pumping heart could be isolated. In 1931, Dr John Gibson came up with a way to divert the blood away from the heart and leave the task of pumping to a machine. If the heart could then be stopped, it could be repaired.[5] After twenty years of experimentation, Gibson was ready to perform such an operation on a young girl with a hole in the heart. She was kept alive for twenty-six minutes by the machine while the hole was repaired. Now, in the 1950s, ambitious surgeons could begin to plan a completely new approach to cardiac medicine: spare-part surgery.

The heart was always going to be the big challenge, which is why the first heart transplant in the world made the South African surgeon who performed it, Dr Christiaan Barnard, a celebrity. The patient was a fifty-five-year-old grocer, Louis Washkansky. After a long wait for a donor, Washkansky was to receive his heart from a young woman who had been killed in a car crash. The operation took place on 3 December 1967, at Groote Schuur Hospital in Cape Town. It lasted nine hours but Washkansky came through it; and the following day he was front-page news all over the world. But, sadly, Washkansky was to live for just eighteen days. Within a few weeks Barnard made a second transplant, this time with greater success, the patient surviving for nineteen months.

By the end of 1968 more than a hundred heart transplants had been performed, but still few patients survived longer than a number of weeks. This poor track record provoked much debate about the ethics of the procedure and, in particular, the cost, as each operation was enormously expensive. But Barnard and a few other surgeons remained convinced the technique could be made to work. Now, with better drugs that suppress the immune system, preventing the donor organ from being rejected, patients are surviving longer, some enjoying well over twenty years of life that they would otherwise not have had.

But as heart transplants became more successful and more acceptable, they were restricted by the availability of organ donors – the number of donors has been in steady decline. There is a happy side to this as the decline can be largely attributed to a reduction in road accident fatalities, not least through the wider use of seat belts.

Donated organs also impose a time factor in the process, which is not desirable. If an artificial heart could be made, the donor problem would disappear. In 1957 the world's first artificial heart was fitted to a dog, which survived for ninety minutes.

The origins of the artificial heart lie with an unlikely amateur inventor, Paul Winchell, who made a living as a voice actor and is familiar as Tigger in *Winnie the Pooh* and Dick Dastardly in *Wacky Races*. In his spare time he patented a mechanical heart, which he later donated to Utah University. Here another pioneer, Robert Jarvik, developed the idea and created the Jarvik 7, an artificial heart that was first fitted to a human being in 1982. The device was very bulky and looked more like a spare part for a washing machine, and it needed external supplies of both electricity and compressed air. But it worked, at least for a while. It was given to a sixty-one-year-old dentist, Barney Clarke, who lived for 112 days. Today the artificial heart has become a marvel of miniaturised precision engineering and materials science, and can grant years of extra life. Where the early heart was a complex mechanism designed to replicate the pulse of the original organ, today's is a simpler pump, powered by a discreet belt-worn battery pack.

But even the artificial heart may have had its day, because soon we may be able to grow new organs organically and engineer them genetically. These will possibly be xenotransplants, which means they will have come from animals, most likely pigs. The hope is that an animal can be bred as a genetic match to the patient, which will prevent the donated organ being rejected and will relieve the recipient of dependence on drugs to suppress the immune system. None of this, however, would have been even vaguely conceivable without a twentieth-century discovery announced in a Cambridge pub on 28 February 1953.

The pub was the Eagle, in Bene't Street, and survives to this day. It was already famous (and still is) as an ancient coach stop and a haunt of Second World War RAF aircrews, who decorated the ceiling with graffiti. It was into this pub that James Watson and Francis Crick burst and proclaimed that they had discovered 'the secret of life'. As it was a pub, this announcement was probably regarded as bar-room wisdom, but for

James Watson and Francis Crick unravel the structure of DNA with the help of what looks like a No.1 Meccano set.

'IT'S FITTING THAT THE TWENTIETH CENTURY SHOULD HAVE DRAWN TO A CLOSE WITH AN IMMACULATE CONCEPTION'

once there was something in it. Because this Nobel Prize-winning pair had untangled the molecular structure of DNA and revealed its famous shape: the double helix.[6]

Around a century before Watson and Crick cracked DNA, Charles Darwin had put forward his radical theory that the diversity of animals and plants on the planet was the result of evolution; if he was right then humankind and apes shared the same ancestors. It was a somewhat controversial idea at the time, and still is in some parts of the American Mid-West. Darwin's process of evolution relied on the principle he called the survival of the fittest, which has led to massive confusion ever since. By 'fittest' he did not mean the strongest, healthiest and fastest running. In this context, the issue is more fitness for purpose; that is, the creatures best suited to survival in their environments are the ones that will end up passing their characteristics on.

A century later Darwin's theory stands up well, but what he could never fully explain was how the different characteristics could be passed on from one generation to the next. Watson and Crick's work was the key, as now the DNA could be understood as an incredibly long genetic code, a blueprint to be found in all living beings. And if they reproduce sexually, like us humans, the blueprint is passed on, mixed up with that of the partner, to create new and unique offspring.

It follows, then, that once the code of human DNA had been cracked it might be possible to identify errors and anomalies in the code that lead to illness, and perhaps even fix them. It also means we all have a unique code that is our genetic signature, which allows us to be identified. This uniqueness has been of enormous importance in the identification of criminals from tiny samples of DNA found at crime scenes.

The implications of unravelling the structure of DNA are mind-boggling, and it is fitting that the twentieth century should have drawn to a close with an immaculate conception, the cloning of an animal. Lamb number 6LL3, or Dolly the sheep, was born on 5 July 1996.[7] For humankind the event was a milestone, a point of no return. There have, technically, always been clones around – they are the natural outcome of children born from monozygotic embryos, identical twins. But Dolly was not the product of a biological chance. She had been manufactured by the genetic engineers at the Roslin Institute in Edinburgh using the cell of an adult sheep.

Now we have to wait for the inevitable, a cloned human being. The obstacles are not simply scientific but moral and ethical ones. Genetic engineering of humans seems to some to represent the final oppression of nature, an unwelcome opportunity to meddle with the characteristics we were granted at conception, or even to create 'designer babies' with the attributes desired by their parents.

It might be worth remembering, though, that as recently as 1978 many people were outraged at the idea that conception could take place in a laboratory rather than in the womb, as nature intended. But the first child born as a result of IVF, Louise Brown, has now grown up and had a child of her own. Louise had been conceived in a test tube and then transferred to

Megan and Morag, two cloned mountain sheep. Even their mother couldn't tell them apart.

her mother Lesley when just eight cells 'old'. Now this technique is commonplace, and accusations that the medical profession is 'playing at God' have rightly subsided.

Cloning and genetic engineering are in a more convoluted moral maze, certainly. But we are still a long way from the day when a midwife holds up a squawking infant and says, 'Congratulations, it's an accountant.'

A SECTION FOR THE ANORAKS

PART 1
SHRINKING THE WORLD

1

Lindbergh was in fact the eighty-second person to fly across the Atlantic. His trip was made eight years after Alcock and Brown's – an eternity in the then twenty-four-year history of powered flight – and in a modern, custom-designed machine.

2

Another early British airline, Instone Air, used a modified version of the Vimy Alcock and Brown had flown. A Vimy was also the first aeroplane to fly from England to Australia.

3

Itself formed from the amalgamation of several small airlines, and the short-haul poor cousin to Imperial. BOAC would, in turn, merge with British European Airways to create what is now British Airways.

4

The airship business was fond of the language of seafaring. Atlantic trips were crossings and departure times were advertised as sailings. The gas-filled envelope was the hull and the crew of R101 included a number of coxswains.

5

Originally, it was intended that the double-deck arrangement would extend for the entire length of the fuselage, but problems with evacuation requirements put paid to the idea. It has been revived in the Airbus A380.

6

Following one of several rows with the French, the British dropped the final 'e' from the Concorde name at the insistence of the prime minister, Harold Macmillan. It was later reinstated by trade secretary Tony Benn on the basis that it stood for Europe, Entente Cordiale and, indeed, England. The Scots were very cross, since they were building part of the aircraft's nose.

7

Since Concorde was an old aircraft by the time of its retirement, it was the only one left on the British Airways fleet to employ dedicated flight engineers. When it was finally withdrawn from service, they were all redundant.

8

Although the industry is eventually repaid in kind. Bentley's Crewe factory was originally a Rolls-Royce aero-engine assembly plant. Jaguar's Castle Bromwich factory once built Spitfires. Most of America's motor industry was given over to the war effort in the Second World War – the Lincoln Motor Corporation, for example, produced the Liberty aero engine – but the consequent re-tooling and capacity excess helped fuel America's post-war car-building boom.

9

An exception being the Wankel rotary engine, championed by NSU in the 1960s and still used today by Mazda. (In the motorcycle world, Suzuki and Norton dabbled in it.)

10

The eastern bloc was fond of two-stroke cars, most notably the Trabant and Wartburg. Since they burned oil as well as petrol, the Communist era, like the age of steam, is remembered for the filth it left behind.

11

Otto's patent for the four-stroke was revoked in Germany in 1886 after it was discovered that a Frenchman, Alphonse Beau de Rochas, had described the idea earlier. This allowed Benz free rein to use it and explains why, while most of the world will still refer to the four-stroke principle as the Otto Cycle, in France it is known as the Cycle de Beau de Rochas.

12

Although the Citroën's brake 'pedal' barely qualified as such – being linked to a high-pressure hydraulic system, it took the form of a mushroom-shaped button on the floor. More of a switch really.

13

These cyclecars were tiny vehicles, usually for two people, designed to lower the cost of motoring by exploiting certain and rather artificial small-car tax advantages in place around Europe. After the Second World War the idea was revived with the microcar, notably those from Heinkel, Messerschmitt and Isetta.

14

VW doesn't deny it. In 1961 the company was forced to pay 3 million DM in compensation to Tatra. The financial hardship suffered by VW as a result is one of the reasons the Beetle stayed in production for so long.

15

To be pedantic about it, it's really fifty pictures a second. But each frame displays only alternate lines of the scan, so a full image appears twenty-five times a second.

16

At least when it was working. High-altitude nuclear tests by the US military, including one performed the day before Telstar's launch, massively increased the level of radiation in the earth's Van Allen Belt and ultimately destroyed the satellite's fragile transistors.

17

Those arrangements of numbers expressed as a power of ten, which could apparently be used to make huge divisions and multiplications easier. Junior scholars of my own generation were not convinced.

18

The job of calculator endured well into the twentieth century. In the design of the R100 and 101 airships, for example, they were employed to work out the stresses and loads on the crafts' metal skeletons. The work took months.

PART 2
TOWARDS INFINITY'S FRONTIER

1
The disastrous Grisson/White/Chaffe mission was retrospectively named as Apollo 1. There were no missions designated as Apollos 2 and 3. Apollo 4 was an unmanned test of the Saturn V. Apollo 5 was an orbital test of the lunar module aboard a Saturn 1B. Apollo 6 was the second unmanned test of the Saturn V and Apollo 7, again using the Saturn 1B, was a manned test of the Apollo command module in earth orbit.

2
This operation, or more specifically its depiction in Ron Howard's film *Apollo 13*, was the inspiration for the TV series *Scrapheap Challenge*, aka *Junkyard Wars* in the US.

3
Deke Slayton, the docking module pilot of the Apollo craft, had been one of the original Mercury astronauts, earmarked to fly in Mercury 7. An irregular heartbeat grounded him, and he was replaced by Scott Carpenter. He was returned to flight duties in 1972, and had to wait three years until the Apollo/Soyuz mission to fulfil his dream of spaceflight.

4
The last discernible transmission from Pioneer 10 was picked up by NASA on 22 January 2003. Although contact with Earth is lost, it will continue to function as an interstellar calling card.

5
Ongoing probe missions, such as Voyagers 1 and 2, have, in fairness, taught us a great deal about the cosmos and in particular our own solar system. Voyager 1 is now the most distant man-made object in space. But in terms of discovering evidence of other life, they have returned with what the non-scientific community would regard as the square root of sod all.

PART 3
THE IMPOSSIBLE CITY

1
There is also an Englishman embroiled in this story, Henry Bessemer (1813–98). His Bessemer process for mass-producing steel of consistent quality was essential to the rise of the skyscraper.

2
The lift itself was not a new idea. Steam-driven lifts had been around since the nineteenth century, and man-powered types long before that.

3
Pilkington, the company, has since developed a self-cleaning glass for tall buildings. Its surface is coated with a photocatalytic substance that breaks down organic dirt, which is then rinsed off by rainwater.

PART 4
THE INVENTION
OF THE TEENAGER

1
Two-way voice communication by radio – for example between air traffic control and aircraft – is strictly known as 'radio telephony', which is why pilots will often talk about 'the RT'.

2
This is why singles feature a slightly raised serrated ring near the centre. It's to stop them slipping when several are piled up on the turntable of a player with autochange. Not many people born after 1980 know that.

3
They also rotated much faster, at up to 500rpm. The 'track', such as it is, on a full-length CD is around three miles long.

4

It was originally known as the Esquire, then the Broadcaster, but there was already a drum set with that name, so after a short period of not having a name at all – the very rare Fender Nocaster – the Telecaster was born.

5

Enrico Piaggio coined the name Vespa (Wasp) for the new scooter. It is not clear whether he thought it looked like one or sounded like one.

6

There had been earlier designs along similar lines, notably the American 1937 Cushman Autoglide.

PART 5
WAR AND THE WORLD

1

The stall speed increases in a turn, and an aeroplane like the BE2 would already be quite close to it during the climb-out from the airfield. The Wright brothers discovered this aerodynamic absolute during their early experiments, but it was not properly understood for some time.

2

There are two types of system for allowing the machine-gun to fire through the propeller arc without hitting it, the interrupter and the synchroniser. In the interrupter, a cam on the engine simply prevents the gun from firing when a blade is in the way. In the synchroniser, a mechanism on the engine operates the action of the weapon, so the rate of fire is related to engine speed. The synchroniser was actually the better solution, since the ammunition of the time was very inconsistent and 'hang fire' rounds could still shoot the following prop blade off in the interrupter scheme.

3

Immelmann devised the Immelmann turn, a climbing and turning manoeuvre used to make a second pass at an overshot adversary, though it was Boelcke who is considered to be the true father of air combat.

4

He didn't. Richtofen was mortally wounded at fairly low level, made a hasty landing in a field, and then died in his cockpit. Debate continues as to who downed him, but it seems likely he was hit by a shot from the ground.

5

The Spitfire was not actually Mitchell's next project; that was the rather lacklustre F7/30 Type 224 fighter. Mitchell did not choose the name Spitfire himself; the War Ministry did. He is reported to have said that it was 'a bloody silly name'.

6

The Stuka (from the German *Sturzkampfflugzeug*, or 'plunging combat aircraft') was also fitted with an air-driven siren, ironically not unlike an air-raid warning siren, to terrorise its victims.

7

These were not the first jet aircraft. Heinkel's He178 flew in August 1939 and the Gloster E28/39 in May 1941.

8

Soviet aircraft were given these 'reporting names' by the western powers, since their original designations were either unknown, unpronounceable, or mere numbers. Some of the more unlikely nicknames include the Fiddler and the Flogger.

9

It was given the strange name of 'landship' because, initially at least, the project was overseen not by the army but by the navy. They, after all, had plenty of experience with machinery.

10

In its early days the role of the landship was kept secret, even from the people building it. They were told that the huge steel shells they were fabricating were a type of water tank. The name stuck.

11

So you might be wondering how Google Earth works. The bigger picture is a satellite image, but the close-up stuff – where you think you can see your car – has been photographed from a low-flying aeroplane.

12

One can be bought easily in the form of an air pistol, made by UMAREX of Germany. Very authentic, too. It'll probably be outlawed soon, so get one while you can.

PART 6
BODILY FUNCTIONS

1

For example, if a weight is being twirled round on the end of a piece of string, at any given point its natural inclination is to continue straight ahead, in accordance with Newton's laws. As it is constantly being pulled into a circular path, it is in effect accelerating towards the centre of the circle, even though it will never get there. This is the cause of so-called centrifugal force.

2

It is believed that Douglas Bader could endure a tighter turn than his fellow pilots without blacking out. Since he had no legs, there was nowhere for his blood to go, so more of it stayed in his head.

3

This technique was originally known as NMR, or Nuclear Magnetic resonance, but the word 'nuclear' was found to alarm patients, so MRI is the more common name.

4

Röntgen used the term X-ray simply to indicate that the nature of the ray was an unknown, and the name stuck. In some countries they are known as Röntgen rays.

5

This was the origin of the now-familiar heart-lung machine, which is essentially a life-support system keeping a patient alive while everything inside the rib-cage becomes still and hence easier to work on.

6

The work of Maurice Wilkins and Rosalind Franklin was important to their discovery. Wilkins was also to receive a Nobel Prize with them; sadly, Rosalind Franklin died.

7

Dolly was named after the busty country star Dolly Parton, as she (the sheep) was cloned from cells taken from mammary glands. Dolly Parton was reported to be quite flattered.

ACKNOWLEDGEMENTS

You can never write a book without the help of some great minds and real supporters, and they all deserve recognition.

At Hodder & Stoughton the person who has been constantly generous with his advice and support is Rupert Lancaster; Juliet Brightmore also deserves a special mention for the terrific work she did finding all the photographs.

Thanks also to Simon, Terry and all at Rose for all their great work on the design of the book, and to the Citroën UK press office for sourcing the car on the front cover.

Those working on the BBC2 series, *James May's 20th Century*, all possessed an abundance of both imagination and energy. In particular, Aidan Laverty, who steered the series with great aplomb, and the team of producers – Helen Thomas, Paul King, Claudine King-Dabbs, Milla Harrison, Dan Walker and Simon Baker, who made sure that it was terrific fun to make. Keeping us honest with regards to the science were Sally Crompton and Tony Nixon of the Open University. And Roly Keating, Emma Swain, Catherine McCarthy, and Mark Jacobs all deserve thanks for giving the project the go ahead.

Finally there are those long-suffering agents, Annie Sweetbaum and Luigi Bonomi, without whom nothing would ever get started.

'*Radio Ga Ga*' Words and Music by Freddie Mercury © 1983, Reproduced by permission of Queen Music Ltd/ EMI Music Publishing Ltd. London WC2H 0QY

TV is the Thing This Year (Phil Medley/William Sanford) Published by Rockland Music, Straylight, Songs of Windswept Pacific & Music of Windswept Used by kind permission from P&P Songs Ltd.

PICTURE
ACKNOWLEDGEMENTS

The Advertising Archives: 40, 44, 56, 120, 142, 171 top, 183, 199, 200, 205, 208. 209. ©AKG-images, London: 35. ©Alamy/L. Zacharie: 116. ©BBC Photo Library: 4-5, 8, 57, 58-59, 72-73, 158-159, 188, 210-211, 258-259, 262, 263, 278. Copyright ©Boeing : 24. ©Cody Images-TRH Pictures: 224 right, 230 top. ©Corbis: 6 (Richard H. Cohen), 18, 29 (Swim Ink 2,LLC), 30 (Minnesota Historical Society), 53 (BBC), 62, 70 (Ed Quinn), 80-81 (Duffin Mcgee/Reuters), 96 right, 98-99, 100 top (Roger Ressmeyer), 111 (Reuters), 112 (Envision), 117 top, (Iconografico, SA), 124 (Joseph Sohm/Visions of America), 126-127, 131 bottom (Lake County Museum), 134, 146 (Schenectady Museum, Hall of Electrical History Foundation), 153 bottom (Underwood & Underwood), 160 (Owen Franken), 168-169 (Underwood & Underwood), 182 (H. Armstrong Roberts), 195 (Lake County Museum), 216, 219 bottom, 221, 248-249 (Vittoriano Rastelli). ©Corbis/Bettmann: 14, 25, 31, 34, 41 top, 48, 68, 78, 82 right, 113, 118, 131 top, 135, 152, 162, 163, 165, 167, 176-177, 178, 191, 207, 212, 223 top left, 224 left, 228, 238, 250. ©Corbis/Hulton-Deutsch Collection: 28, 50, 54-55, 189, 232-233. ©Mary Evans Picture Library: 36-37, 180-181. Courtesy of Ford Motor Company Limited: 47. ©Getty Images: 15 top (Hulton Archive), 19 (Central Press), 20 (Photo Lambert), 22 (John Chillingworth), 39 (Keystone), 42-43 (Three Lions), 60 (Ted Wood/Aurora), 66-67 (Fox Photos), 85 (NASA), 86 (Space Frontiers), 114-115 (Steven Peters/Riser), 128 (Dwight Franklin/Museum of the City of New York), 132-133 (Horace Abrahams), 136 (Jeff Spielman/Iconica), 138-139 (Mieko Kanasashi/Photonica), 140 (STR/AFP), 148-149 (Tim Graham), 150 (General Photographic Agency), 154 (Harry Todd/Fox Photos), 156 (Reg Birkett), 170 (FPG), 186 (Metronome), 187 top (Dave Benett), 190 (Jack Robinson), 192 (John Kobal Foundation), 202 (Bert Hardy), 213 (Hulton Archive), 220 top (MPI), bottom (Hulton Archive), 223 top right (Fox Photos), 229 (Keystone), 231 (US Air Force), 245 (Hulton Archive), 264-265 (Stephen Frink/Science Faction), 266 (AFP). ©Getty Images/Time & Life Pictures: 32-33 (Charles E. Steinheimer), 41 bottom (Hugo Jaeger/Timepix), 74 (Ralph Morse), 76 (Bob Gomel), 84 (Bill bridges), 90-91 (NASA), 93 (NASA), 172-173 (Terence Spencer), 174 (Henry Groskinsky), 179 (Nina Leen), 187 bottom

(Robin Platzer), 196-197 (Terence Spencer), 198 (Terence Spencer), 242-243 (Mansell), 247 (Eliot Elisofon). Courtesy of Hagley Museum and Library, Wilmington, Delaware, USA: 206. Courtesy of Honda Motor Europe: 200, 201. ©Imperial War Museum, London: 215 (Q45786), 217 left (Q69650), right (Q65882), 218 (Q69222), 219 top (Q12051), 225 (C3546A), 226-227 (C5422), 230 bottom (FKD 600), 239 top (Q70931), 240 right (Q6284). 241 (Q6432). ©Magnum Photos/Stuart Franklin: 251. Garrett Augustus Morgan: 153 top (Patent Published as US 1475024, 1923). ©The National Archives: 252 (HS 7/28), 255 (HS 7/30), 256 (HS 7/49). The Ohio State University Radio Observatory and the North American AstroPhysical Observatory (NAAPO): 122 top. ©PA Photos/John Walton: 269. Ellis Parrinder/PCP Agency: ii-iii, viii. ©popperfoto.com: 9, 222, 223 bottom left, 239 bottom, 240 left. ©Reuters: 26-27 (Jonathan Bainbridge ASA), 234 (Ian Waldie), 236 (Magnus Johansson). ©Rex Features: 194. ©Science & Society Picture Library: 51 (NMPFT), 64 (Bletchley Park Trust).©Science & Society Picture Library/Science Museum: 10, 11, 12-13, 16, 17, 143, 267, 268. ©Science Photo Library: 77 (NASA), 82 left (Ria Novosti), 83 (Detlev Van Ravenswaay), 89 (NASA), 94 (Ria Novosti), 96 left (NASA), 97 (NASA), 100 bottom (NASA), 102 top (Detlev Van Ravenswaay), bottom (Ria Novosti), 103 (Ria Novosti), 104-105 (NASA), 106 (JVZ), 107 (NASA), 108-109 (NASA/KSC), 110 (NASA), 117 bottom, 122 bottom (NASA), 123 (A. Nota/NASA/ESA/STScI), 144-145 (M-SAT Ltd), 260 (NASA), 270, 273 (Bluestone), 274 (Hank Morgan), 275 (A. Barrington Brown), 277 (James King-Holmes). Courtesy of Sony United Kingdom Limited: 171, 183. Reproduced by permission from the Vickers Archives held at the Cambridge University Library, and with the permission of Rolls-Royce plc:

15 bottom. Courtesy of Yamaha Motor (UK) Limited: 200. ©Gunter Zint/Redferns: 184.

FURTHER READING

If you'd like to dig a little deeper into the technology of the twentieth century there are a lot of great books available.

Here are some of our favourites.

For general books on inventions try:
The Mammoth Book of Great Inventions edited by James Dyson and Robert Uhlig; published by Robinson, it covers 2.6 million years BC to the present day.
Inventing the 20th Century by Stephen Van Dulken, published by the British Library, which concentrates on the patents.
Giant Leaps, an unusual collaboration between the *Sun* and the Science Museum, published by Boxtree, is great fun.
A History of Invention by Trevor I. Williams published by Little, Brown and Co. goes a lot deeper.
Creating the Twentieth Century by Vaclav Smil covers 1867 to 1914 and is published by Oxford University Press.
The Shock of the Old: Technology in Global History Since 1900 by David Edgerton, published by Oxford University Press, is great for a completely different perspective.

For Concorde, take a look at *Concorde* by the photographer Wolfgang Tillmans, published by Walther Konig.

For space, and in particular NASA's Mercury astronauts in the 1960s, if you haven't read *The Right Stuff* by Tom Wolfe, published by Farrar, Straus and Giroux, then it's time you did, and buy the DVD of the feature film of the same name while you are about it.

For the Apollo missions Andrew Smith's *Moondust*, published by Bloomsbury, is terrific.

If you are into guitars then *Classic Guitars of the Fifties*, published by Balafon, has great information and pictures. (There's another on guitars of the sixties.)

On spying, *The Ultimate Spy Book* by H. Keith Melton, published by DK, is fascinating.
Secret Agent's Handbook of Special Devices introduced by Mark Seaman, published by the Public Record Office, has the SOE catalogue in it and is wonderful.

For British boffins, read *Backroom Boys* by Francis Spufford, published by Faber and Faber.

And Bill Bryson's *A Short History of Nearly Everything*, published by Black Swan, is outstanding.

INDEX